5th International Conference on Modern Technologies for Non-Destructive Testing 2016

IOP Conference Series: Materials Science and Engineering
Volume 189

Tomsk, Russia
3 – 8 October 2016

ISBN: 978-1-5108-3959-5
ISSN: 1757-8981

Printed from e-media with permission by:

Curran Associates, Inc.
57 Morehouse Lane
Red Hook, NY 12571

Some format issues inherent in the e-media version may also appear in this print version.

Printed by Curran Associates, Inc. (2017)

For permission requests, please contact the Institute of Physics
at the address below.

Institute of Physics
Dirac House, Temple Back
Bristol BS1 6BE UK

Phone: 44 1 17 929 7481
Fax: 44 1 17 920 0979

techtracking@iop.org

Additional copies of this publication are available from:

Curran Associates, Inc.
57 Morehouse Lane
Red Hook, NY 12571 USA
Phone: 845-758-0400
Fax: 845-758-2633
Email: curran@proceedings.com
Web: www.proceedings.com

Table of contents

Volume 189

5th International Conference: Modern Technologies For Non-Destructive Testing

3–8 October 2016, Tomsk, Russian Federation

Accepted papers received: 31 March 2017
Published online: 18 April 2017

Preface

011001
OPEN ACCESS
5th International Conference: Modern Technologies For Non-Destructive Testing

011002
OPEN ACCESS
Peer review statement

Papers

012001
OPEN ACCESS
Research of Conductive Materials by Multifrequency Measuring System on the Basis of Eddy Current Transducers

S Dmitriev, A Ishkov, A Katasonov, V Malikov, A Sagalakov, L Shevtsova and K Ekkerdt.....1

012002
OPEN ACCESS
Monitoring and Quality Control of Diesel Fraction Production Process

E V Frantsina, N S Belinskaya, A S Lutsenko and N V Popova.....6

012003
OPEN ACCESS
A New Method for Control of the Efficiency of Gear Reducers

K E Kozlov, A V Egorov and V N Belogusev.....11

012004
OPEN ACCESS
Fuse Selection for the Two-Stage Explosive Type Switches

I O Muravlev, M A Surkov, E V Tarasov and N F Uvarov.....17

012005
OPEN ACCESS
Day-Ahead Short-Term Forecasting Electricity Load via Approximation

R N Khamitov, A S Gritsay, D A Tyunkov and G E Sinitsin.....23

012006
OPEN ACCESS
Electromagnetic Method for Exogenetic Geodynamic Elements Mapping in Permafrost Environment

V F Gordeev, S Y Malyshkov and V I Polyvach.....30

012007
OPEN ACCESS
Synthetic Reference Materials Based on Polymer Films for the Control of Welding Fumes Composition

O V Kuznetsova, A N Kuznetsova and L A Begunova.....34

012008
OPEN ACCESS
Method for Prediction of the Power Output from Photovoltaic Power Plant under Actual Operating Conditions

S G Obukhov, I A Plotnikov, O A Surzhikova and K D Savkin.....38

012009
OPEN ACCESS
Using the Semiconductors Materials of InSb-ZnTe System in Sensors for Gas Control

E G Shubenkova.....44

012010
OPEN ACCESS
Injection Process Control of the Well at the Hydrodynamic Research of Coalbed

I G Odnokopylov, O V Galtseva, I Yu Krasnov, A O Smirnov, M S Karpov, O A Surzhikova, V V Kuznetsov and J Li.....48

012011
OPEN ACCESS
Simulation of Industrial Wastewater Treatment from the Suspended Impurities into the Flooded Waste Mining Workings

L Bondareva, Yu Zakharov and A Goudov.....54

012012
OPEN ACCESS
Determination of the Optimal Position of Pendulums of an Active Self-balancing Device

G R Ziyakaev, O A Kazakova, V V Yankov and O P Ivkina.....59

012013
OPEN ACCESS
Bottom Ash Waste Used in Different Construction Materials

G G Volokitin, N K Skripnikova, O G Volokitin, A V Lutsenko, V V Shekhovtsov, V A Litvinova and M A Semenovykh.....64

012014
OPEN ACCESS
Process of Formation of Vacuum Polymer Films in the HF Discharge Reactor

P A Luchnikov and O A Surzhikova.....69

012015
OPEN ACCESS
Algorithm and Software for Landslide Slopes Stability Estimation with Online Very Low Frequency Monitoring

V F Gordeev, M M Kabanov and S N Kapustin.....74

012016
OPEN ACCESS
X-ray Tomographic System Behavior Prediction Based on a Mathematical Model

S S Baus, L A Redko and M N Yanushevskaya.....79

012017
OPEN ACCESS
Statistical Degradation Models for Reliability Analysis in Non-Destructive Testing

E S Chetvertakova and E V Chimitova.....85

012018
OPEN ACCESS
Transformer Oil Dielectric Strength in the Contact Gap of the Explosive Arc-Extinguishing Device

I O Muravlev, M A Surkov, E V Tarasov, Kh Tsoggerel and N F Uvarov.....91

012019
OPEN ACCESS
Non-Contact Technique for Determining the Mechanical Stress in thin Films on Wafers by Profiler

N A Djuzhev, A A Dedkova, E E Gusev, M A Makhiboroda and P Y Glagolev.....97

012020
OPEN ACCESS
The application of the Methods of Cathodoluminescence and Photoluminescence for Non-destructive Testing of AlGaN Heterostructures

Y A Kontsevoy, E Ph Pevtsov, R A Khmelnitskiy, Y V Kolkovskiy, V V Gruzdov and S A Malykhin.....103

012021
OPEN ACCESS
X-Ray Diffraction Analysis of Bottom Ash Waste after Plasma Treatment

G Volokitin, Yu Abzaev, N Skripnikova, O Volokitin and V Shekhovtsov.....107

012022
OPEN ACCESS
Calculation of Optimal Geometrical Magnification and Spatial Resolution of Betatron Tomograph

Y Zhong, S V Chaklov and V B Trinh.....112

012023
OPEN ACCESS
Estimation of the Lithospheric Component Share in the Earth Natural Pulsed Electromagnetic Field Structure

S Y Malyshkov, V F Gordeev, V I Polyvach, S G Shtalin and K N Pustovalov.....117

012024
OPEN ACCESS
The Selection of Computed Tomography Scanning Schemes for Lengthy Symmetric Objects

V B Trinh, Y Zhong and S P Osipov.....123

012025
OPEN ACCESS
The Level of Impact Noise and Threshold Characteristics Nonlinear Assessment Signal Parameters

Y S Timoshenkova.....128

012026
OPEN ACCESS
Assessment of the Charging Policy in Energy Efficiency of the Enterprise

E A Shutov, T E Turukina and T S Anisimov.....133

012027
OPEN ACCESS
Deformable Polymer Dielectric Films in Phase Light Modulators

P A Luchnikov, V V Vetrova, N P Zubkov, O A Surzhikova and N I Chernova.....139

012028
OPEN ACCESS
Method of Calculating the Correction Factors for Cable Dimensioning in Smart Grids

M Simutkin, V Tuzikova, J Tlusty, V Tulsky and Z Muller.....144

012029
OPEN ACCESS
Risk Assessment for Stonecutting Enterprises

A J Aleksandrova and S S Timofeeva.....150

012030
OPEN ACCESS
Soil ph Control in the Mobile Corrosion Monitoring

T A Radenkov, S V Romanenko, V A Kolpakov, O E Merzlyakov, V A Kabanov and A G Kagirov.....156

012031
OPEN ACCESS
Structural and Phase Changes in the System Al-Si-Ti-B, Synthesized Using the Electron-Ion-Plasma Treatment Method

A A Klopotov, E A Petrikova, Yu F Ivanov, A D Teresov, N N Cherenda, V V Uglov and N A Tsvetkov.....161

012032
OPEN ACCESS
Method and Apparatus of Measuring Velocity and Sound Attenuation Coefficient in Bulk Materials Based on the Analysis of the Structure of Sound-Insulation Materials on the Basis of Perlite

B I Kapranov and A P Mashanov.....167

012033
OPEN ACCESS
Focused Ion Beam Methods for Research and Control of HEMT Fabrication

E Ph Pevtsov, A V Bespalov, T A Demenkova and P A Luchnikov.....172

012034
OPEN ACCESS
Phytoremediation of Excavation in Eastern Siberia

S S Timofeeva and T M Zhmurova.....178

012035
OPEN ACCESS
High-Energy X-ray Radiation Registration Model

S P Osipov, O S Osipov and K V Ignatenko.....183

IOP Conf. Series: Materials Science and Engineering **189** (2017) 011001 doi:10.1088/1757-899X/189/1/011001

This issue of the journal is dedicated to the scientific and applied research presented at the V International Conference "Modern Technologies for Non-Destructive Testing". The conference was held on October 3–8, 2016 in Tomsk (Russia).

The conference brought together experts from different countries and organizations who received a great opportunity to share knowledge in sessions, to initiate discussions on topics of interest to all participants.

The aim of the Conference was to support young talented people, to involve pupils, students, post-graduate students and young scientists in the solution of actual scientific problems of non-destructive testing, quality management and new educational technologies.

The main goals of the event were exchange of information and discussion of topics related to non-destructive testing, quality management, new educational technologies and widespread implementation of the obtained results.

The key topics of the conference were advanced technologies in non-destructive testing, effective quality management systems, new educational technologies in the field of control and management, and management concepts of modern economy.

The Conference Organizer is The Department of Physical Methods of Non-Destructive Testing of the Institute of Non-Destructive Testing of the National Research Tomsk Polytechnic University.

The Organizing Committee of the V International Conference "Modern Technologies for Non-Destructive Testing" wishes to thank the IOP Conference Series Materials Science and Engineering-Team for collaboration.

IOP Conf. Series: Materials Science and Engineering **189** (2017) 011002 doi:10.1088/1757-899X/189/1/011002

Peer review statement

All papers published in this volume of *IOP Conference Series: Materials Science and Engineering* have been peer reviewed through processes administered by the proceedings Editors. Reviews were conducted by expert referees to the professional and scientific standards expected of a proceedings journal published by IOP Publishing.

IOP Conf. Series: Materials Science and Engineering **189** (2017) 012001 doi:10.1088/1757-899X/189/1/012001

Research of Conductive Materials by Multifrequency Measuring System on the Basis of Eddy Current Transducers

S Dmitriev[1], A Ishkov[2], A Katasonov[3], V Malikov[4], A Sagalakov[2], L Shevtsova[5], and K Ekkerdt[3]

[1]Assistant Professor, Altay State University, Barnaul, Russia
[2]Professor, Altay State Agrarian University, Barnaul, Russia
[3]Student, Altay State University, Barnaul, Russia
[4]Graduate student, Teacher, Altay State University, Barnaul, Russia
[5]Assistent, Novosibirsk State Technical University, Novosibirsk, Russia

E-mail: osys11@gmail.com

Abstract. The measuring system based on subminiature eddy-current transducers has been developed to carry out local investigations of aluminum-magnesium alloy plates for flaws. The Delianna filter has been modified to allow the significant increase of signal-to-noise ratio. The transducer has been tested on a number of aluminum-magnesium alloy plates with flaws. The article presents data on the relationship of eddy-current transducer response to the presence of flaws in alloys as hidden slots at signal frequencies comprised between 300÷700 Hz on an exciting winding

1. Introduction

Nondestructive eddy-current monitoring may be used to find defects in any electrically conducting material. Where necessary, such measurements permit the investigation of each part produced at a plant. Duralumin and aluminum–agnesium alloys are widely used in manufacturing, especially in the aerospace industry and elsewhere. Thanks to their combination of strength and lightness, they are also suitable for high-speed trains, such as the Shinkansen trains (Japan).

Duralumin is also used in electrical engineering, in the chemical industry, in food processing, in ventilation systems, in radioengineering, and in construction. For example, D16AM alloy is used at extremely low temperatures, while D16T duralumin is used in shipbuilding, on account of its plasticity.

The quality control of these alloys and their products is an urgent problem; investigations are making progress in this direction.

2. Problem Statement

The analysis of a recent investigations points to the miniaturization tendency for eddy-current transducers. Transducers with a size of 5×5 mm and a 0.15 mm diameter of the wire have been designed [1]. However, they do not provide the required penetration depth and localization of the magnetic field that are necessary for local measurements in different nonuniform media. Ferrite magnetic field concentrators are often used to increase the area of the magnetic field. A similar design provides an advantage that is related to the absence of the scatter of eddy currents [2]. In addition, a

2.5 mm penetration depth is attained. L. Barbato et al. [3] scanned two aluminum plates with a model flaw in the center and tested cracks between the plates. The diameter of the measuring winding was 7 mm. The scanning was performed at 1 and 5 kHz. In this case, the penetration depth of eddy currents into the studied plates at the above_mentioned frequencies was 3.82 and 1.71 mm.

3. Research Questions

There are different well-known constructions of superimposed vortex-current transducers (VCT), whose working surface has either a plane or hemispherical form. Such kind of a surface provides a satisfactory contact of the VCT with the surface under control, but the tension quantity, sent to the VCT, greatly depends on the curvature of the controlled surface. The edge effect influences the work of the VCT considerably. This effect makes it impossible to control the details of complicated configuration and of small sizes. To solve these problems the VCT is often provided with an extra magnetic conductor. One of its terminals has a shape of a truncated cone. This solution of the problem has only one disadvantage. Despite the increased localization of the magnetic flux, the construction of the magnetic conductor has become considerably more complicated. It makes the measuring accuracy worse, because the output signal of the VCT depends to a great extent on the interaction of two magnetic conductors, which can influence the intensification of the vortex-current field in a rather unpredictable way. These currents contain the information about the controlled object.

Producing such devices demands making a special technological line, due to that the final price of the device will increase significantly. To make the price lower we came to the conclusion that we should replace the expensive hardware blocks with the software for personal computers (PC).

4. Purpose of the Study

The goal of this project is to provide a new type of eddy-current measurement system, capable of producing destructive testing of conductive materials at different depths to determine the depth and size of defects in real time.

5. Research Methods

We have developed a subminiature eddy-current transducer for local monitoring of the physical parameters in aluminum-alloy plates and weld seams [3]. In contrast to existing sensors, it permits local measurements with sections measuring a few microns, to depths of the order of 5 mm. The electrical conductivity of the alloy is directly measurable, and its distribution over the sample surface and thickness may be readily established.

We need to develop an adequate model of the response of eddy-current transducer s of plate type, which are sensitive to many variables and permit reproduction of the voltage hodograph at small values of β0. We have plotted hodographs illustrating the influence of various parameters on the induced voltage on the basis of the proposed model [5–7].

The exciting winding (diameter D_1 = 0.12–0.13 mm) of the subminiature eddy-current transducer consists of ten turns of copper wire (cross-sectional area 5 μm^2). The measuring winding (diameter 0.05–0.08 mm) consists of 130 turns of copper wire (cross-sectional area 20 μm^2).

To minimize the influence of the exciting winding on the final signal, the circuit includes a compensation winding [8–10] that consists of 20 turns of copper wire (cross-sectional area 5 μm^2), connected to the measuring winding in such a way that the voltage of the exciting winding is subtracted from the result. The windings are wrapped around a core of 2000NMZ ferrite (relative magnetic permeability μ_{max} = 500 or else of 81NMA alloy annealed by a special method (if greater localization of the magnetic field is required). The core consists of a tetrahedral pyramid (height 1 mm), with a square base (sides of 0.2 mm). The measuring winding rests on the points of the pyramid, which improves the localization of the magnetic field. Such transducer s permit effective localization of the magnetic field, so that defects as small as 250 μm may be detected. In addition, the magnetic field penetrates into the sample to a considerable depth when working at relatively low frequencies.

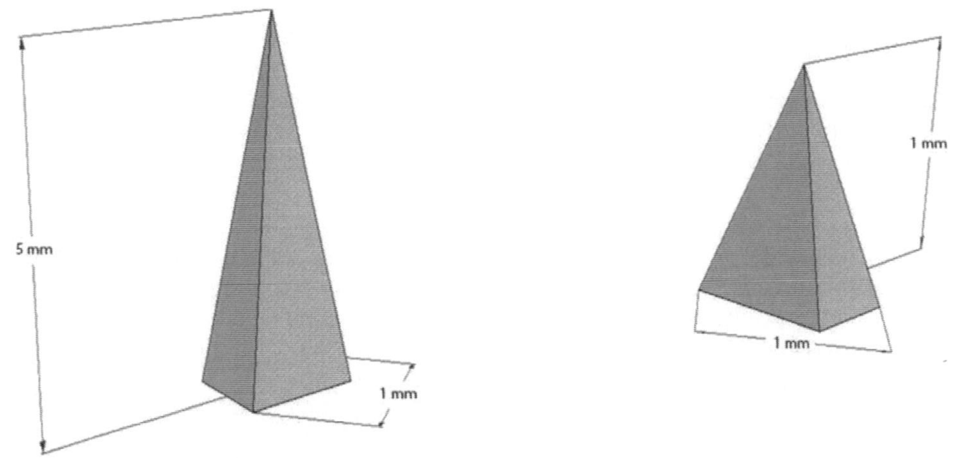

Figure 1. The cores of different sizes.

The laid on VCT consists of a pyramid shaped core. Such kind of a shape was achieved mainly in two ways. The first one is as follows: thin triangles (0.1 mm) were turned out of the alloy 81NMA. After that the plates were put together and their upper parts were ground off. Such kind of an approach has its disadvantages. For example, the tops were ground off unevenly; as a result the magnetic field was rather weak.

The second way to make a pyramid shaped core was grinding off the thin triangles and their matching into a pyramid.

Wire coils were placed on the pyramid shaped cores. The coils were impregnated with compound. The temperature of impregnation was 200 degrees Celsius, the diameter of the wire was 1.5×10^{-6} meters. The measuring winding was placed at the end of the top of the pyramid, the diameter of the winding was 0.05 mm, the number of winds varied from 100 up to 200. In the middle of the pyramid there was an actuating winding that consisted of one wind. The compensating winding was situated on the mobile frame and included 100 wires. The mobile frame enables it to move freely along the oxide core from the bottom up to the actuating winding.

We have made 10 VCTs. To subtract the directed current from the actuating winding into the measuring one we have retrofitted all the devices. We made different VCTs because we wanted to get the magnetic field of different tension. That was why the cores differed from each other in size. The correlation between the diagonal of the bottom and the edge of the pyramid varied from 1:1 to 1:10 (Figure 1).

The VTCs constructed on the basis of the cores, having similar correlation between the diagonal of the base (400 mkm) and the length of the rib (4 mm), have been calibrated on semi-conductors with the certain conductivity. Thus, the VCTs have identical geometric parameters of the cores and the same number of the winds of the actuating (1 wind), measuring (200 winds) and compensation (200±40) windings.

The measuring system, which is based on a miniature eddy current transducer, operates as follows. The software of the personal computer controls the operation of the generator, which produces a train of rectangular voltage pulses with the repetition rate f1 that is necessary for the operation of the eddy-current transducers. The voltage pulses are transmitted from the generator output to two series integrators. They are then directed to the input of the power amplifier. From the amplifier output the voltage pulses arrive at the exciting inductance coils of the eddy-current transducers. The difference of the output voltages of the measuring coils of the transducers contains information on the structural heterogeneities of the tested object that is located in the effective area of the eddy-current transducers. It is detected and amplified in a special microphone amplifier.

The signal arrives at the amplitude detector after the transmission through two series high quality low frequency filters and two series selective amplifiers. The signal is then transmitted through an analog-to-digital converter to a personal computer. Due to the simultaneous control of the generated signal frequency at the exciting coil and the cut off frequency of the filtering system and the selective amplification, the useful signal, which contains information on the electric conductivity distribution inside the object, in particular, on possible flaws of the object, is detected. The control program allows one to change the operating frequency of the measuring system so that the signal that is received from the measuring winding is reliably recorded.

6. Findings

Samples with model defects were prepared in order to assess the maximum depth and the linear dimensions of the defects, for the detection of them it is advisable to use the eddy-current inspection method.

The samples were plates of Al–Mg alloy. The thickness of the first plate (sample №1) was 5.5 mm. The plate contained three defects in the form of slits 1 mm thick at the depths of 1, 3 and 4 mm (Figure 2).

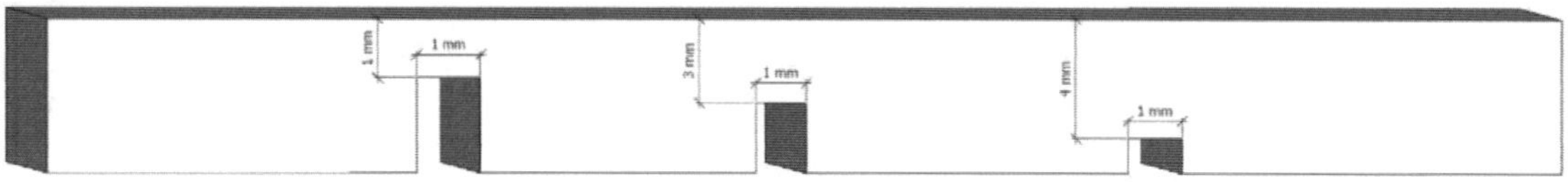

Figure 2. The appearance of the sample № 1.

The thickness of the second plate (sample №2) was 5.5 mm. The plate contained six defects in the form of slits 0.25 mm thick at the depths of 1, 2, 3, 4, 5 and 5.3 mm (Figure 3).

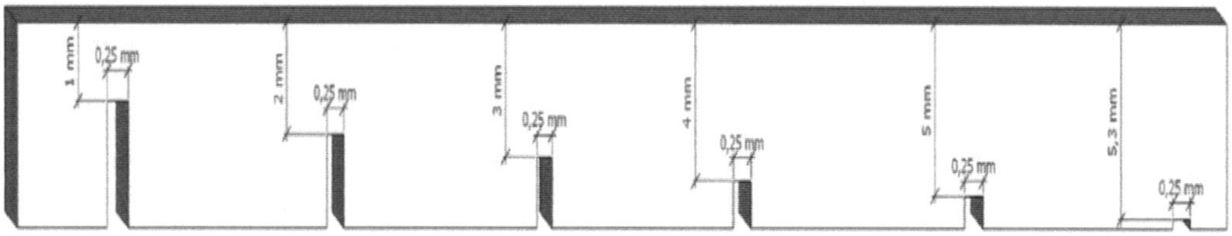

Figure 3. The appearance of the sample № 2.

In order to determine the sensitivity of the sensor to the defects at the depth of the metal, scanning of defect-free side of the sample was performed. During the experiments with the first plate the magnitude of the inserted voltage to the exciting winding of the transducer was 2 V.

The results of the defectoscopy of the first plate, thick at the frequency of 500 Hz and 2 V signal amplitude allowed to detect all three slits by means of signal amplitude drop (Figure 4). Signal amplitude drop in the first defect was 0.75 V, in the second one - 0.2 V, in the third one - 0.1 V.

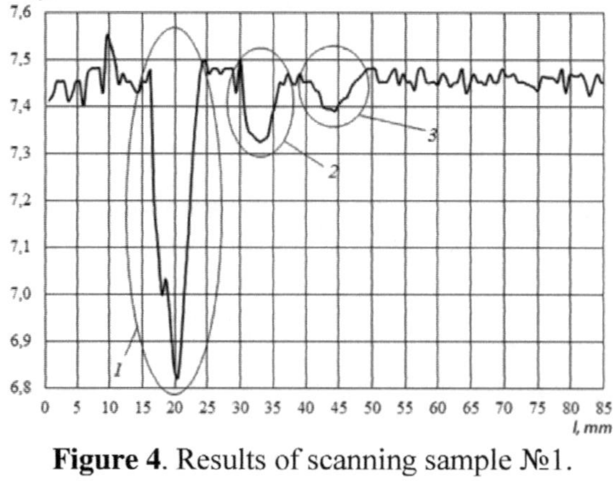

Figure 4. Results of scanning sample №1.

The results of defectoscopy of the second plate at the frequency of 500 Hz and 3 V signal amplitude allowed to reveal five defects (Figure 5). The signal amplitude drop at the first defect was 2.5 V, at the second defect – 1 V, at the third one – 0.4 V, at the fourth one - 0.2 V, at the fifth defect - 0.1 V. The changes in the response signal when passing over the sixth defect were not fixed. The experimental results show the effectiveness of the developed measuring system to find defects of the thickness from 0.25 mm at the depth of 5 mm.

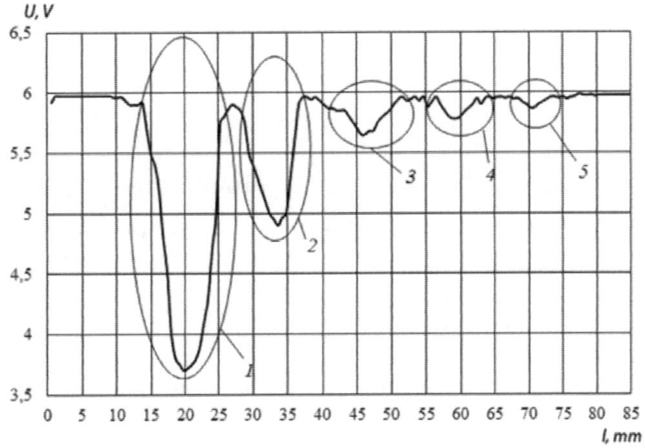

Figure 5. Results of scanning sample №2.

7. Conclusion

The developed measurement system based on eddy-current subminiature transducers, allows greater localization of electromagnetic field in comparison with previously known similar systems. The pyramidal shape of the core, the system of band pass filters and selective amplification allow to reduce interference and achieve significant penetration depth of eddy currents in the object under study. Defects scanning in aluminum alloys can detect defects with linear dimensions of the order of 100 microns at the depth of 5 mm. The developed software allows to automate measurements and make rapid changes in the operating frequency of the device

References

[1] Prance R J *et al* 2000 *Sens. Actuators, A* **85(1–3)** 361–364 DOI: 10.1016/S0924-4247(00)00375-7
[2] Prance R J *et al* 2003 *Rev. Sci. Instrum.* **74 (8)** 3735–3739 DOI: 10.1063/1.1590745
[3] Semenov V S *et al* 2003 *Bull. of the Tomsk State University* **278** 1–7 (in Russian)
[4] Barbato L *et al* 2015 *IEEE Trans. Magn.* **51(7:1–1)** DOI: 10.1109/TMAG.2015.2406765
[5] Rocha T J *et al* 2015 *Sensors and Actuators*, A **228** 55–61 DOI: 10.1016/j.sna.2015.02.004
[6] Litvinenko A A 2004 RF Patent 2231287 *Byull. Izobret.* **14** (in Russian)
[7] Lee K H *et al* 2012 *IEEE Transactions on Magnetics* **48(11)** 3965–3968 DOI: 10.1109/TMAG.2012.2202643
[8] Klyuev V V 2003 *Non-Destructive Testing* (Moscow: Mashinostroeniye) **2(1–2)** (in Russian)
[9] Malikov VN *et al* 2016 *Proc. of MEACS'15* 15805223 DOI: 10.1109/MEACS.2015.7414951
[10] Dmitriev S F *et al* 2016 *IOP Conf. Ser.: Mater. Sci. Eng.* **116(1)** 012013 DOI: 10.1088/1757-899X/116/1/012013

IOP Conf. Series: Materials Science and Engineering **189** (2017) 012002 doi:10.1088/1757-899X/189/1/012002

Monitoring and Quality Control of Diesel Fraction Production Process

E V Frantsina[1], N S Belinskaya[2], A S Lutsenko[3], and N V Popova[4]

[1]Junior Research Fellow, National Research Tomsk Polytechnic University, Tomsk, Russia
[2]Assistant, National Research Tomsk Polytechnic University, Tomsk, Russia
[3]Post-graduate Student, National Research Tomsk Polytechnic University, Tomsk, Russia
[4]Graduate Student, National Research Tomsk Polytechnic University, Tomsk, Russia

E-mail: evf@tpu.ru

Abstract. In this work the mathematical model of diesel fraction and atmospheric gasoil catalytic dewaxing process has been developed. Also the pattern of applying the created model to solving such problems as monitoring and quality control of diesel fraction production in the catalytic dewaxing process. It has been represented that to meet such challenges, the model should take into consideration thermodynamic and kinetic laws of hydrocarbon conversion on the catalyst surface, and instability factors that are specified by catalyst deactivation. The developed model allows controlling the quality of obtained diesel fraction depending on feed and temperature regime in the reactor. The value of model calculation absolute error does not exceed 2%, which corroborates the adequacy of the model to actual process. The computations using the model have shown that to provide the desired product yield (not less than 40% wt. of overall yield of the unit products) of programmed quality (cold filtering plugging point not higher than minus 34°C for winter diesel fuels and not lower than minus 40°C for arctic ones) at long-time catalyst operation (during 4 years), it is necessary to sustain the reactor temperature at the average level of 19°C higher than when working with fresh catalyst. This must be done to compensate catalyst activity loss due to its deactivation.

1. Introduction

Industrial catalytic processes of petroleum refining sector represent complex chemical technology systems that require qualified control and operation. Optimization of technological process is of a high difficulty level due to its instability presented by continual changing of feed fraction compositions, variations of operating modes, proceeding chemical reactions and catalyst deactivation. Simultaneous introduction of all mentioned factors is impossible in real terms. Consequently, the challenge of forecasting the operation modes of complex commercial units within the instable conditions is multitask and labor intensive.

This challenge is successfully met applying mathematical models, developed in terms of fundamental laws of hydrocarbon conversion on the catalyst surface, considering the instability factors and mutual influence of every of adjacent processes being conducted in a single chemical technology system [1, 2, 3]. Application of these models allows forecasting the behavior of complex chemical technology system at the constantly changing conditions of the process [4, 5].

2. Experimental

The technology of catalytic dewaxing is successfully applied to obtain winter, summer and arctic grades of diesel fuel with improved cold flow properties [6]. This is a complex chemical technology system which involves the interconnected stages of hydrotreating, hydrodewaxing, stabilization and fractionation [1, 2].

During the process a stable naphta, diesel fraction with improved low-temperature properties, and residue with a boiling point of above 340 degrees of Celsius are obtained.

The process proceeds in three reactors in hydrogen environment. The first two reactors are hydrotreating reactors with a nickel-molibdenium catalyst, the third one is a dewaxing reactor with nickel catalyst. Process temperature is 350–400°C. The reactions of hydrocracking, hydroisomerization, hydrogenation of unsaturated hydrocarbons, hydrogenation of mercaptanes followed by hydrogen sulfide formation, aromatization and coke formation take place.

After reactor section the products enters the stabilization column, where the light hydrocarbons, hydrogen sulfide and water are eliminated. The stable hydrogenate subsequently flows to the rectification section, where it is fractionated into products, such as gasoline and diesel fraction.

Based on experimental data from the industrial unit, also with the application of quantum chemistry method, realized within the frames of Gaussian software, thermodynamic reaction parameters have been defined, and formalized reaction network in the dewaxing reactor has been made (Figure 1).

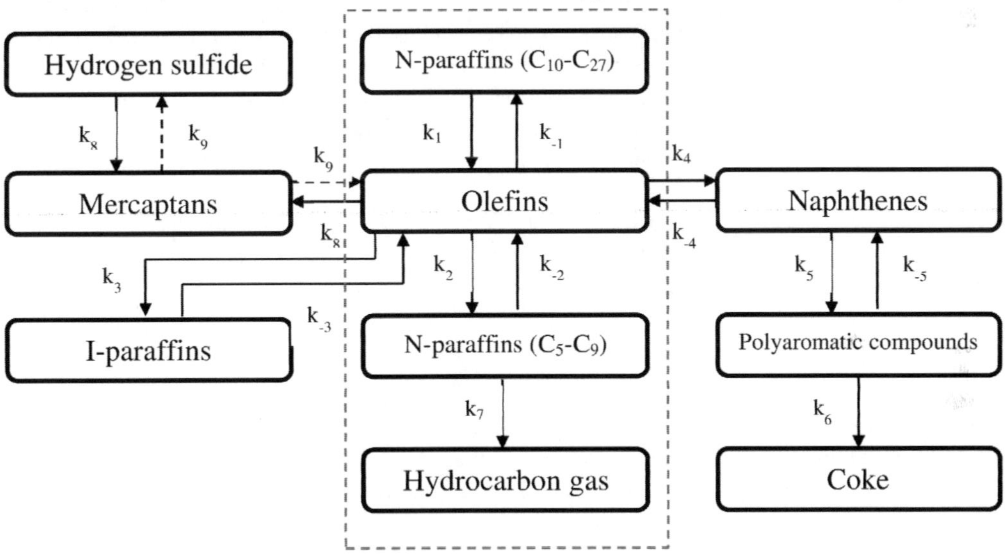

Figure 1. Formalized scheme of catalytic dewaxing reactions.

Primary reactions of the process are the reaction of hydrocracking of high-molecular unbranched alkanes and the reaction of dehydrogenation of low-molecular alkanes with further isomerization, through the stage of olefin formation [7, 8]. The main secondary reaction is coke formation.

According to the formalized reaction scheme the kinetic description of the process has been made. The kinetic model represents the system of differential equations of changing reacting agent concentrations depending on feed residence time.

To maintain the optimal catalyst activity during the middle distillate refining, aimed to obtain the required product volume of specified quality, it is necessary to correct the operating modes of the dewaxing reactor. Thus, to decrease the catalyst deactivation and to sustain optimal activity, the feed is mixed with hydrogen-containing gas (HCG) and the reactor temperature is increased. In this case hydrogen gas reacts with formed polyaromatic hydrocarbons, which may invoke the process of coke formation, hydrogenates these compounds and leads to intensification of olefin and mercaptane hydrogenation.

On the other hand, excess amount of hydrogen may result in intensified hydrocracking reactions followed by reduction of desired diesel fraction yield. Therefore, the main task of controlling the dewaxing catalyst deactivation process is determination of optimal HCG/feed ratio under the conditions of changing flow rate, feed composition and temperature inside the reactor.

To control the dewaxing catalyst deactivation process under the conditions of changing feed composition the deactivation function has been added to the mathematical model of the dewaxing process [9-11]. The mathematical model represents the system of differential equations of mass and heat balance. Mathematical model considering deactivation process is shown below:

$$\begin{cases} G\cdot\dfrac{\partial C_i}{\partial z}+G\cdot\dfrac{\partial C_i}{\partial V}=\sum\limits_{j=1}^{m} a_j\cdot W_j \\ G\cdot\dfrac{\partial T}{\partial z}+G\cdot\dfrac{\partial T}{\partial V}=\dfrac{1}{\rho\cdot C_p^{mix}}\sum\limits_{j=1}^{m} Q_j\cdot a_j\cdot W_j \end{cases}, \quad (1)$$

Initial conditions: $z = 0$: $C_i = C_{i,0}$; $T = T_0$; $V=0$: $C_i = C_{i,0}$; $T = T_0$, where z is a volume of the refined feed from the moment of fresh catalyst charge, [m^3]; G is the feed flow rate, [m^3/h]; $z = G\cdot t$ (t is a catalyst operating time from the moment of fresh catalyst charge, [h]); C_i is a concentration of i-component, [mole/l]; V is a catalyst bed volume, [m^3]; a_j is a catalyst activity in j-reaction; ρ is a mixture density, [kg/m^3]; C_p^{mix} is a specific heat capacity of the mixture, [J/(kg·K)]; Q_j is a heat effect of j-reaction, [J/mole]; T is a temperature, [K]; W_j is a j-reaction rate, [mole/(l·s)]; m is a number of the reactions.

This model takes into account catalyst deactivation [12, 13]. Catalyst activity can be determined as a proportion of reaction rate constant at any time to the reaction rate constant at the initial moment of the time when the fresh catalyst had been charged:

$$a_j=\frac{k_{j,t}}{k_{j,0}}, \quad (2)$$

where $k_{j,0}$ is a j-reaction rate constant at the initial moment of time (with the fresh catalyst); $k_{j,t}$ is a j-reaction rate constant at the current moment of time. In should be notices that catalyst activity depends on coke concentration on the catalyst surface, and can be defined as an equation:

$$a_j=A_j\cdot e^{-\alpha_j\cdot C_K}, \quad (3)$$

where A_j, α_j are the deactivation coefficients; C_c is a coke concentration, [% wt].

3. Results and Considerations

Created mathematical model for controlling the quality of diesel fraction production process has been checked for adequacy by comparing the calculated data with experimental data from the industrial unit. The error values of calculation of the main hydrocarbon groups, found in diesel fraction, are presented in Table 1. The error does not exceed 2% which corresponds to the error of chromatographic analysis, usually used to determine hydrocarbon content. The model is suitable for the process and can be applied for identification of the process laws, forecast calculation conduction and optimization, providing resource efficient production by control of unit operation modes, product quality and catalyst deactivation process.

Table 1. Calculated and experimental data comparison.

Product (experimental), [% wt.]	Product (calculation), [% wt.]	Error (absolute), [% wt.]
	n-paraffins C_{10}–C_{27}	
10.17	11.18	1.01
13.54	11.85	1.69
10.08	10.19	0.11
9.32	9.48	0.16
	i-paraffins	
26.46	26.40	0.06
22.95	24.1	1.15
24.66	24.82	0.16
23.14	25.67	2.53
	Olefins	
1.74	1.99	0.25
2.63	2.58	0.05
3.07	2.35	0.72
2.25	2.35	0.10
	Polyaromatic compounds	
20.53	20.06	0.47
19.90	20.09	0.19
18.53	19.23	0.70
19.30	19.73	0.43

It is well known that during the catalytic dewaxing process the catalyst deactivation takes place due to the reaction of coke formation and sulfur-poisoned acid and metal catalyst sites [13-14]. Application of developed model, which considers deactivation function, helped to prove the fact that during the period from 2012 to 2016 slight decrease of diesel fraction yield from 90 to 44 [% wt.] has been occurring, and the gasoline fraction yield has been increasing from 8 to 20% at the same time (Table 2).

This shows attenuation of hydrogenation reactions which leads to reduction of diesel fractions yield, and intensification of cracking reactions which leads to enhancement of light fractions yield. It should be noticed that increase of the unconverted residue yield (fraction with a boiling point of more than 340°C) (Figure 2) indicates the reduced catalyst activity and decreased feed conversion.

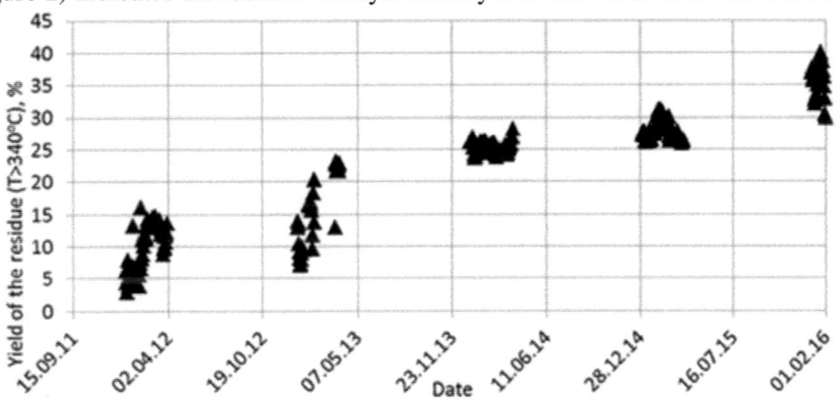

Figure 2. Yield of the residue (T>340°C) in the dewaxing process during the period of 2012-2016 y.

Table 2. Yield of all fractions in the dewaxing process in the period of 2012-2016.

Year	2012	2013	2014	2015	2016
Yield of diesel fraction, [%]	74 – 90	63 – 88	52 – 65	54 – 60	44 – 52
Yield of gasoline fraction, [%]	8 – 10	10 – 12	15 –18	18 – 20	20 – 21
Yield of residue fraction >340 [^{o}C]	5 – 15	10 – 25	25 – 30	30 –35	35 – 40

The computations using the model have shown that to provide the desired product yield (not less than 40 [% wt.] of overall yield of the unit products) of programmed quality (cold filtering plugging point not higher than minus 34°C for winter diesel fuels and not lower than minus 40°C for arctic ones) at long-time catalyst operation (during 4 years), it is necessary to sustain the reactor temperature at the average level of 19°C (Table 3) higher than when working with fresh catalyst. This must be done to compensate catalyst activity loss due to its deactivation, which amounted to 32% by the last year of catalyst operation (2016 y.).

Table 3. Computation of the deactivation compensating temperature.

Date	Yield of fraction 240-340 (DF), [%]	CFPP, [^{o}C]	T_{in} R-3 (actual), [^{o}C]	T_{in} R-3 (comp.), [^{o}C]	ΔT, [^{o}C]
25.01.2016	41,4	-34	341	318	23
27.01.2016	41,8	-37	352	332	20
29.01.2016	41,5	-44	360	345	15
05.02.2016	43,0	-41	369	350	19

4. Summary

To meet the challenges associated with monitoring and quality control of diesel fraction production in the catalytic dewaxing process the mathematical model, developed considering thermodynamic and kinetic laws of hydrocarbon conversion on the catalyst surface, and instability factors that are specified by catalyst deactivation, can be used.

The proposed mathematical model allows defining the optimal operating modes of the reactor block, such as process temperatures and feed flow, and controlling production of the product of programmed quality, which implies hydrocarbon composition and yield of diesel fraction, cold filtering plugging point, and can be used for solving the tasks of increasing resource efficiency of large tonnage industrial catalytic processes.

References

[1] Belinskaya N S *et al* 2016 *IFOST: New Materials and Technologies* 211–215
[2] Frantsina E V *et al* 2016 AIP Conference Proceedings **1772** 060017 DOI: 10.1063/1.4964597
[3] Erdal Aydın *et al* 2015 *Comp.and Chem.Eng.* **82** 44–54
[4] Galtseva O V *and* Bordunov S V 2015 *IOP Conf. Series: Materials Science and Engineering* **81(1)** 112010 DOI: 10.1088/1757-899X/81/1/012062
[5] Belinskaya N S and Frantsina E V *et al* 2016 *IOP Conf. Series: Earth and Environmental Science* **43(1)** 1–6 DOI: 10.1088/1755-1315/43/1/012062
[6] Rakoczy R A *and* Morse P M 2013 *Hydrocarbon Processing* 67–69
[7] Castano P *et al* 2010 *Catalysis Today* **150** 363–367
[8] Rosetti I *et al* 2009 *Chem. Engin. J.* **154** 295–301
[9] Kohli K *et al* 2016 *Fuel* **175** 264–273
[10] Krasnov I *et al* 2015 *IOP Conf. Ser.: Mater. Sci. Eng.* 93 012061 DOI: 10.1088/1757-899X/93/1/012061
[11] Iuriev I Y *et al* 2015 *IOP Conf. Ser.: Mater. Sci. Eng.* 71 012022 DOI: 10.1088/1757-899X/71/1/012022
[12] Elizalde I *and* Ancheyta J 2014 *Fuel Processing Technology* **123** 114–121
[13] Surzhikova O A 2015 *IOP Conf. Ser.: Mater. Sci. Eng.* 81(1) 012098 DOI: 10.1088/1757-899X/81/1/012098
[14] Kumbilieva K *and* L Petrov 2011 *Chin. J. Catal.* **32** 51–59

A New Method for Control of the Efficiency of Gear Reducers

K E Kozlov[1], A V Egorov[2], and V N Belogusev[3]

[1]Associate Professor, Volga State University of Technology, Yoshkar-Ola, Russia
[2]Associate Professor, Mari State University, Yoshkar-Ola, Russia
[3]Associate Professor, Volga State University of Technology, Yoshkar-Ola, Russia

E-mail: konstantin.k-e@yandex.ru

Abstract. This article proposes a new method to control the energy efficiency of gear reducers. The method allows evaluating the friction losses in a drive motor, drive motor bearing assemblies, and toothing both at the stage of control of the finished product and in the course of its operation, maintenance, and repair. The proposed method, unlike currently used methods for control of the efficiency of gear reducers, allows determining the friction losses without the use of strain measurement, which requires calibration of tensometric sensors and expensive equipment. The method is based on the idea of invariability of mechanical characteristics of the induction motor at constant voltage, resistance of windings, and mains frequency, regardless of the driven inertia mass. This paper presents experimental results which verify the theoretical predictions. The proposed method can be implemented in the procedure of acceptance test at the companies that manufacture gear reducers, thereby assess their effectiveness and the level of degradation processes that significantly affect the service life of the research object. The method can be implemented both with universal and with specialized hardware and software complexes. At that, both an increment of the inertia moment and acceleration time of a gear reducer may serve as a performance criterion.

1. Introduction

The problem of improving the quality of products of modern engineering is complex. Its resolution depends on a number of purposes including increasing the efficiency of machines in a wide range of possible modes of their operation. The solution of the problem is associated with the development of methods and instruments for their testing, control, and diagnostics. The present level of development of methods and instruments for control of the efficiency of gear reducers does not allow comprehensively evaluating the technical condition of the equipment in a wide range of possible modes of operation [1–6]. In order to register a change of friction losses rate in gear transmissions, it is most often applied the strain measurement that requires a calibration of strain sensors and the use of an expensive equipment [7–10]. At that, the efficiency is registered with a low measurement rate, which is caused by the time required to recover elastodeformed state of a strain element [11].

Thus, on the one hand, there is a need to resolve the problem of increasing the efficiency of gear reducers; on the other hand, there are no adequate methods and instruments to solve the problem.

Analysis of the works carried out on the basis of the Closed Joint-Stock Company "Research Institute of Introscopy of MSIA "Spectrum" (together with the Departments of Power Engineering, Mechanical Engineering, Mechanics and Control Processes) headed by the academician of RAS, Vladimir V. Kluyev [12], proves that inertia methods for control, testing, and diagnostics of gear reducers have not been developed due to the main unresolved problem associated with the inertial testing - the problem of determining the normalized with respect to the power take-off axis effective moment of inertia of rotating parts.

This article gives the solution of the problem hampering the development of inertial methods for control, testing, and diagnostics of gear reducers.

In order to solve the stated problems, we need to find a new approach to control the efficiency of gear reducers. This approach should not have the disadvantages, which other known methods have (the need for calibration of strain sensors, high cost of equipment, and the complexity of the measurement process). The measurement accuracy obtained using the new method is to meet the requirements of modern engineering and to be comparable to the accuracy of measurements obtained with the help of well-known and approved methods of measurement.

The purpose of this paper is to develop a cost-effective, highly accurate, and less labor consuming method for control of the efficiency of gear reducers based on the use of additional bodies with the known moments of inertia.

In order to achieve this purpose, we assigned the following objectives:

- To develop the theoretical justification of the method for the efficiency control of gear reducers.
- To develop the theoretical justification of the method for evaluating the rate of losses caused by the friction in bearing assemblies and toothing.
- To prove the theoretical predictions experimentally.

2. Methodology

Let us consider the proposed method for control of the efficiency of gear reducers with the use of a gear reducer operating as part of an induction electric drive. Figure 1 shows a scheme of a single-stage bevel gear speed reducer connected to an induction motor.

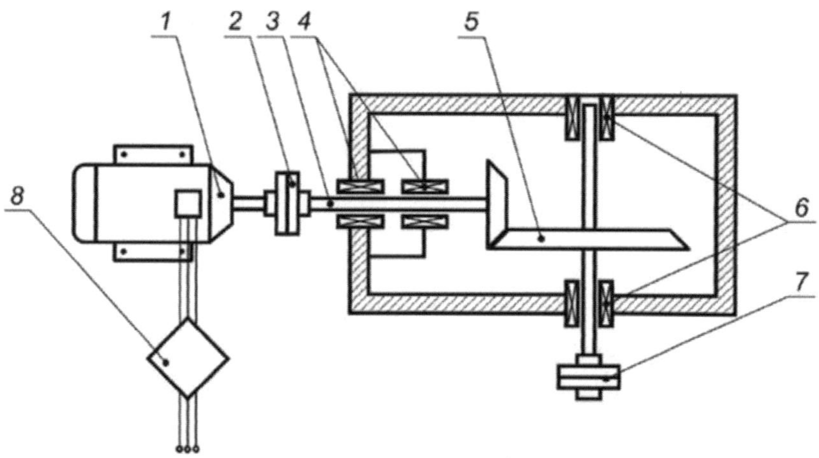

Figure 1. Scheme of a gear reducer driven by an induction motor; where 1 is an induction motor, 2 is a safety clutch, 3 is a high-speed (input) gear-shaft, 4 are bearings supporting the high-speed (input) shaft, 5 is a low-speed (output) shaft together with a gear wheel, 6 are bearings supporting a low-speed (output) shaft, 7 is a safety clutch of a reducer, 8 is a frequency converter (inverter).

Let us consider the implementation of the developed method for an induction motor with a safety clutch.

Knowing the calculated or experimentally determined moment of inertia of a safety clutch 2, J_{scim}, it is possible to determine the moment of inertia of the induction motor rotor 1, $J_{im}(\omega)$, taking into account the friction losses in bearings in accordance with the scheme in Figure 2.

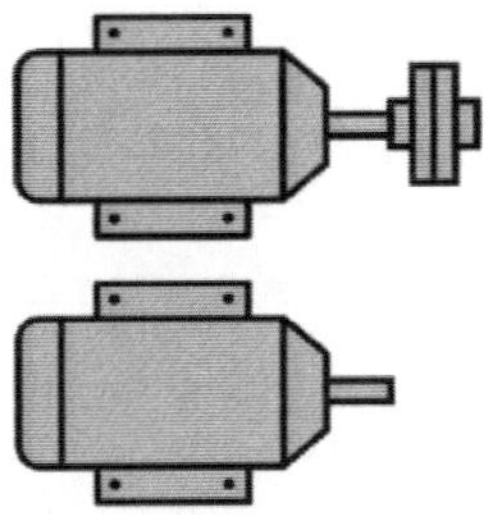

Figure 2. Scheme for determining the moment of inertia of a rotor of an induction motor taking into account the friction losses in its supports.

First, it is necessary to determine the angular accelerations of the rotor with the safety clutch installed on an output shaft of the motor, $\varepsilon_1(\omega)$, and then - without it, $\varepsilon_2(\omega)$. Considering that the speeding up has actually been at idle mode, the products of the moments of inertia and the angular accelerations during the first and the second accelerations can be equated to each other:

$$(J_{scim} + J_{im}(\omega)) \cdot \varepsilon_1(\omega) = J_{im}(\omega) \cdot \varepsilon_2(\omega). \tag{1}$$

Based on Equation 1, we can calculate the moment of inertia of the rotor taking into account the friction losses in bearings as follows:

$$J_{im}(\omega) = \frac{J_{scim} \cdot \varepsilon_1(\omega)}{\varepsilon_2(\omega) - \varepsilon_1(\omega)}. \tag{2}$$

Based on the described method, the methodology for control of the efficiency of a gear reducer is implemented according to the following steps:

1. Determination of the effective capacity depending on the angular velocity of a rotor of an electric motor by the developed method.
2. Determination of the moment of inertia of an input gear shaft together with a gear wheel mounted on it by torsional oscillation.
3. Determination of the frictional power losses in input shaft bearing supports depending on the angular speed of a rotor by the developed method.
4. Determination of the moment of inertia of an output shaft of a gear reducer together with a gear wheel mounted on it by torsional oscillation.
5. Determination of the frictional power losses in output shaft bearing supports depending on the angular speed of a rotor of an electric motor by the developed method.
6. Determination of the total frictional power loss in bearings of input and output shafts of a gear reducer depending on the angular velocity of a rotor of an electric motor.
7. Determination of the frictional power losses in toothing depending on the angular velocity of a rotor of an electric motor.

In the last step, the reducer is totally assembled and the induction motor 1 is connected to the high-speed gear-shaft 3 through the safety clutch 2.

Then we determine the normalized with respect to the rotor axis moment of inertia of the rotating parts of the entire drive according to the conventional method based on the law of conservation of kinetic energy without taking into account the losses in the gear mechanism:

$$J_{norm0}(\omega) \cdot \frac{\omega^2}{2} = \left[J_{im}(\omega) + J_{scim} + J_{gs}(\omega)\right] \cdot \frac{\omega^2}{2} + \left[J_{lss}(\omega) + J_{scred}\right] \cdot \frac{\omega^2}{2 \cdot i^2}. \tag{3}$$

$$J_{norm0}(\omega) = J_{im}(\omega) + J_{scim} + J_{gs}(\omega) + \frac{J_{lss}(\omega)}{i^2} + \frac{J_{scred}}{i^2}, \quad (4)$$

where i is the gear ratio; $J_{gs}(\omega)$ is an array of values of the moment of inertia of the gear-shaft taking into account the friction losses in supports; $J_{lss}(\omega)$ is an array of values of the moment of inertia of the low-speed shaft taking into account losses in bearings supporting the low-speed shaft; J_{scred} is the moment of inertia of a safety clutch 7.

Then the induction motor 1 speeds up and an array of values of the angular accelerations, $\varepsilon_3(\omega)$, is measured. The torque, developed by the electric motor, is determined as follows:

$$M(\omega) = (J_{norm0}(\omega) + J_{loss.tooth}(\omega)) \cdot \varepsilon_3(\omega). \quad (5)$$

The torque developed on an output of the motor safety clutch 2 can be determined as follows:

$$M(\omega) = (J_{im}(\omega) + J_{scim}) \cdot \varepsilon_1 \cdot (\omega). \quad (6)$$

Equating Equations 5 and 6, we determine the normalized with respect to the rotor axis moment of inertia of losses in toothing as follows:

$$J_{loss.tooth}(\omega) = (J_{im}(\omega) + J_{scim}) \cdot \frac{\varepsilon_1(\omega)}{\varepsilon_3(\omega)} - J_{norm0}(\omega). \quad (7)$$

The total normalized with respect to the rotor axis moment of inertia of rotating parts of the gear reducer together with the induction motor can be determined as follows:

$$J_{norm}(\omega) = J_{im}(\omega) + J_{scim} + J_{gs}(\omega) + J_{loss.tooth}(\omega) + \frac{J_{lss}(\omega)}{i^2} + \frac{J_{scred}}{i^2}. \quad (8)$$

Let us write Equation 8 in terms of elementary components:

$$J_{norm}(\omega) = J_{rim} + J_{blim}(\omega) + J_{scim} + J_{gs} + J_{bllss}(\omega) + J_{loss.tooth}(\omega) + \\ + \frac{J_{lss}}{i^2} + \frac{J_{bllss}(\omega)}{i^2} + \frac{J_{scred}}{i^2}. \quad , (9)$$

where, J_{rim} is the moment of inertia of the rotor; $J_{bllss}(\omega)$ is the losses in the bearings 4 supporting the low-speed shaft 5.

The power developed by the separate parts of the electric gear drive is determined as follows:

$$N(\omega) = J_{rim} \cdot \varepsilon_3(\omega) \cdot \omega + J_{blim}(\omega) \cdot \varepsilon_3(\omega) \cdot \omega + J_{scim} \cdot \varepsilon_3(\omega) \cdot \omega + \\ + J_{gs} \cdot \varepsilon_3(\omega) \cdot \omega + J_{bllss}(\omega) \cdot \varepsilon_3(\omega) \cdot \omega + J_{loss.tooth}(\omega) \cdot \varepsilon_3(\omega) \cdot \omega + \\ + \frac{J_{lss}}{i^2} \cdot \varepsilon_3(\omega) \cdot \omega + \frac{J_{bllss}(\omega)}{i^2} \cdot \varepsilon_3(\omega) \cdot \omega + \frac{J_{scred}}{i^2} \cdot \varepsilon_3(\omega) \cdot \omega. \quad (10)$$

Knowing an array of values of the active electric power supplied to the input of the induction motor, $N_{im}(\omega)$, knowing $N(\omega)$, knowing the power of dissipation in windings of the motor, $N_{diss}(\omega)$, we can determine the power of loss for the eddy current and hysteresis in the induction motor:

$$N_{ec,hys}(\omega) = N_{im}(\omega) - N(\omega) - N_{diss}(\omega). \quad (11)$$

3. Experimental validation of a new method

In order to substantiate the reliability of the values obtained with the developed method, we carried out the following experiment. We compared the values of the torque obtained with the developed method with those obtained with the torque sensor M40-100 during an induction motor speeding-up.

The experiment was carried out according to the scheme shown in Figure 3.

For measuring the angular acceleration, we used an encoder E40HBP, which sends an analog signal corresponding by its dynamic range to the measuring channel of a registration data unit. The personal computer with installed software provides mathematical treatment of a data array and calculation of the angular velocities and the average values of the accelerations of a rotating motor. Based on the values of the angular accelerations obtained during the motor speeding up, both with a reference body and without it, the moment of inertia is computed. To minimize voltage fluctuation, which influences the measurement error, voltage stabilizer Saturn SNE-O-10 (11 kW, 50 A) was used.

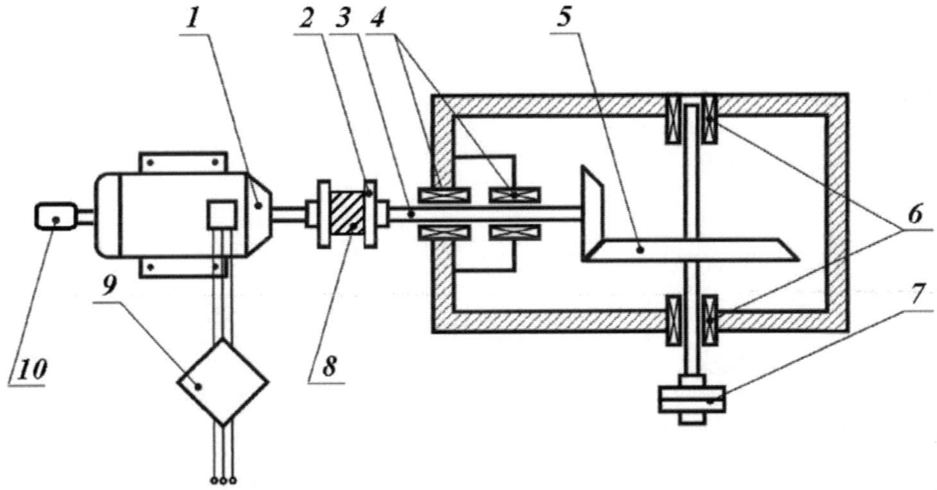

Figure 3. Scheme for control of the electric gear drive torque with a torque sensor M40-100, where 1 is an induction motor, 2 is a safety clutch, 3 is a high-speed gear-shaft, 4 are bearings supporting a high-speed shaft, 5 is a low-speed shaft together with a gear wheel, 6 are bearings supporting a low-speed shaft, 7 is a safety clutch of a reducer, 8 is a torque sensor M40-100, 9 is a frequency converter (inverter), 10 is an encoder E40HBP.

In the application of both methods 10 measurements were carried out, blunders were eliminated, random errors were determined, and the average torque values were found. Table 1 shows the values obtained.

Table 1. The comparison of the average torque values within selected speed range obtained with the novel method (NM) and using the torque sensor (TS).

Variable	Number of revolutions, [rev/min]						
	250–400	400–500	500–600	600–700	700–800	800–850	850–950
M_{NM}, N·m	12.362	14.537	16.259	17.947	19.845	20.895	19.182
M_{TS}, N·m	12.212	14.289	16.298	17.828	19.694	20.686	18.992
δ, %	1.21	1.71	-0.23	0.66	0.76	1.00	0.99

4. Conclusion

The experimental results show that the convergence of the measurements obtained with the developed method and with the torque sensor varies within 1.7%, which proves the validity of the values obtained with the developed method.

Having determined an array of values of $J_{norm}(\omega)$ at the stage of a technical control department of the manufacturer of gear reducers, having determined the acceptable level of this parameter during operation of a gear reducer, and equipping a gear drive with hardware and software system, the user has the ability to promptly respond to the deterioration of the energy performance of the drive. On the basis of measuring $J_{norm}(\omega)$, it is possible to carry out maintenance and repair work to the extent necessary to maintain the efficiency at a high level. With the implementation of the novel method for control of the efficiency of gear reducers, it is possible to achieve high efficiency of a reduction gear for the entire life cycle.

The methodology for determining the efficiency of gear reducers with a fixed transmission ratio having a number of steps more than two is similar to the developed one.

The developed method for control of the efficiency of gear reducers can be used for research purposes during the engineering development, for testing newly manufactured products in the industrial stages, and for testing gear reducers in the technical operation stage during the entire life cycle.

Acknowledgements

The reported study was funded by Ministry of Education and Science of the Russian Federation according to the research project No. 02.G25.31.0204.

References

[1] GOST 7217-87 2003 *Rotating electrical machines. Asynchronus motors. Test methods* (Moscow: Standardinform) (in Russian)
[2] Kubo K *et al* M 1986 *Tribol. Int.* **19(6)** 312–317 DOI: 10.1016/0301-679X(86)90038-1
[3] Laus L P *et al* 2012 *Mechanism and Machine Theory* **52** 296–325 DOI: 10.1016/j.mechmachtheory.2012.01.011
[4] Lyashenko M V *et al* 2016 *Proc. of ICIE-2016* **150** 1233–1238 DOI: 10.1016/j.proeng.2016.07.124
[5] Praveenkumar T *et al* 2014 *Proc. of GCMM'2014* **97** 2092–2098 DOI: 10.1016/j.proeng.2014.12.452
[6] Shen Zh *et al* 2012 *Measurement* **45(1)** 30–40 DOI: 10.1016/j.measurement.2011.10.008
[7] Irimescu A *et al* 2011 *Int. J. Automotive Technology* **12(4)** 555–559 DOI: 10.1007/s12239–011–0065–1
[8] Rybalchenko Yu I 1981 *Magnetoelastic torque sensors* (Moscow: Mashinostroeniye) (in Russian)
[9] Raghuwanshi N K *and* Parey A 2016 *Measurement* **86** 266–275 DOI: 10.1016/j.measurement.2016.03.001
[10] Goszczak J 2016 *IOP Conf. Ser.: Mater. Sci. Eng.* **148** 012041 DOI: 10.1088/1757-899X/148/1/012041
[11] Egorov A *et al* 2015 *Int. J. Appl. Eng. Res.* **10(17)** 37685–37691
[12] Kluyev V V *et al* 2001 *Measurement, control, testing and diagnostics* (Moscow: Mashinostroeniye) **III–VII** (in Russian)

IOP Conf. Series: Materials Science and Engineering **189** (2017) 012004 doi:10.1088/1757-899X/189/1/012004

Fuse Selection for the Two-Stage Explosive Type Switches

I O Muravlev[1], M A Surkov[1], E V Tarasov[1], and N F Uvarov[2, 3]

[1]Associate Professor, National Research Tomsk Polytechnic University, Tomsk, Russian Federation
[2]Professor, Institute of Solid State Chemistry and Mechanochemistry of the Siberian Branch of the RAS, Novosibirsk, Russian Federation
[3]Professor, Novosibirsk State Technical University, Novosibirsk, Russian Federation

E-mail: tarasovev54@yandex.ru

Abstract. In the two-level explosive switch destruction of a delay happens in the form of electric explosion. Criteria of similarity of electric explosion in transformer oil are defined. The challenge of protecting the power electrical equipment from short circuit currents is still urgent, especially with the growth of unit capacity. Is required to reduce the tripping time as much as possible, and limit the amplitude of the fault current, that is very important for saving of working capacity of life-support systems. This is particularly important when operating in remote stand-alone power supply systems with a high share of renewable energy, working through the inverter transducers, as well as inverter-type diesel generators. The explosive breakers copes well with these requirements. High-speed flow of transformer oil and high pressure provides formation rate of a contact gap of 20 - 100 m/s. In these conditions there is as a rapid increase in voltage on the discontinuity, and recovery of electric strength (Ures) after current interruption.

1. Introduction

In a two- stage disconnection method of high current circuit [1-6] the fuse is connected parallel to the power contacts. This solution allows to dramatically increase the length of the disconnection arc and reduce the limiting level of fault current. When the explosive breaker used as the first step, the destruction of a parallel wire takes the form of an electrical explosion. In turn, the electric explosion of the wire generates short voltage spike on the scattering of metal products (Figure 1). In the further process of fault current disconnection the high heat absorption capacity of burst absorber provides an increase of the voltage on disconnection arc and creates the conditions for the extinction of disconnection arc. Resulting at the fuse wire explosion the peak voltage must be matched with the air-gap electric strength of the first stage. Otherwise the electric breakdown between the moving contacts and further device failure are possible. The peak voltage can not be taken into account for the arc chute gap size choosing, and therefore the wire size (length and cross section) as well. Getting characteristics of an exploding wire was the purpose of the present study (Figure 1).

2. Theoretical Study

Experiments were carried out in the LC circuit (GIT (generator of pulse currents)) simulating the fault current.

In the investigated setting range, successful current changeover to the second stage and the subsequent disconnection was observed in the wire cross-sections $(0.03\text{-}0.2)\cdot 10^{-6}$ M^2. Larger cross-sections did not provide a sufficient voltage growth rate on the air-gap and lengthen out disconnection process, smaller cross-sections lead to electric breakdown in the intercontact gap of the first stage. The current densities at the moment of wire destruction were within (0.3-1)×10^{-11} *A/M*^{2}, which proves existence of an electrical explosion mode [7, 8].

Detailed examination of the voltage waveform on the fuse wire shows a typical electric explosion (Figure 1b). After the current is transferred in the wire, at the first time its value is determined by wave impedance of the loop because of small wire impedance. The wire absorbs energy, is melted (t_1), and further heated in a liquid state until the moment (t_2). At this moment the step of heating the wire comes to an end and the rapid expansion of the wire is begun.

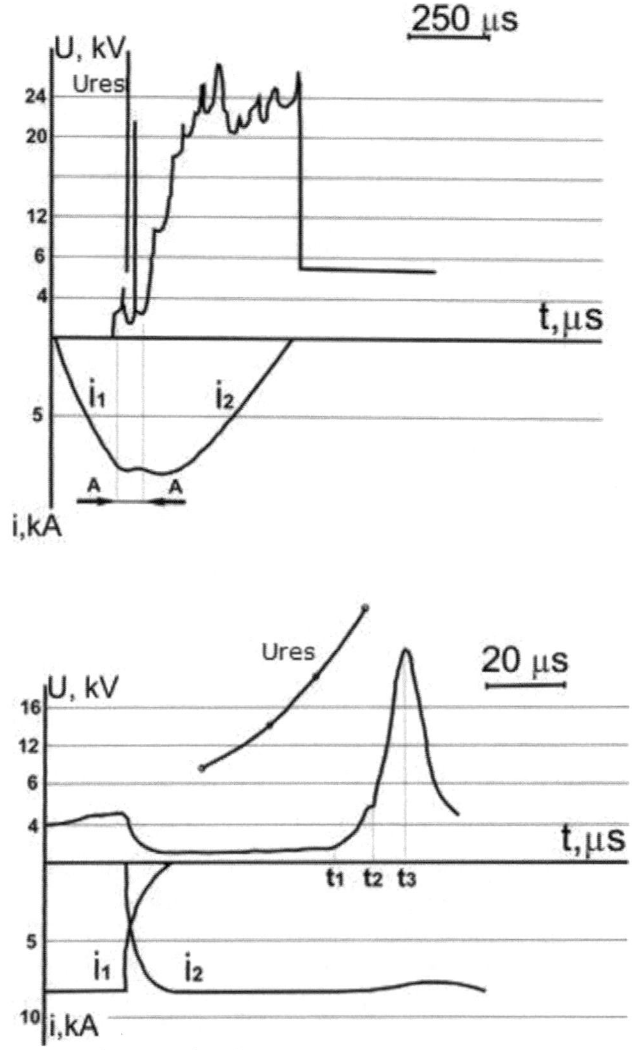

Figure 1. Pictures for current interruption by two-stage explosive switch. AA is switching current to the second stage of two-stage explosive type switches.

The wire loses metal conductivity, its resistance increases by several orders of magnitude, the current in the circuit is changed, the inductance generates a voltage peak on the resistance of fuse wire This stage is wire explosion actually. It is less studied as at the same time there are several process and its relative influence on the explosion process depends on the initial conditions. At the time of (t_3) the gap breakdown on a wire occurs and that initiates the arc.

For the best understanding the phenomena in the disconnection gap is enough to simplistically divide the whole process at the stage of the heating and destruction. At the heating stage wire cross-section is practically unchanged. Heating time from the beginning of current flow to the voltage peak is connected with current density j by ratio, which has been called the "specific action integral" [7].

$$\int_0^{t_n} j^2(t) \times dt = h \quad (1)$$

The value h depends mostly on wire material and selected with possibly lower resistivity for the current switching conditions. According to the degree of increase g metals are located as follows: Au, Ag, Cu, Al. For economic reasons, we choose the last two. Moreover, a review of research on exploding wires [4] shows that the most studied and give the best characteristics at explosion the copper conductors of circular cross-section, it have h≈2×10^{17} A^2c/M^4 for a wide range of current densities. In the case the fuse wire operates in the second stage of a circuit breaker and the current through the wire will change only slightly at the heating stage, $t_н$ is defined as

$$t_n = h/j^2 \quad (2)$$

At the destruction stage there is a sharp increase of the wire resistance and transformation of its material into a vapor-drip blend. The way of destruction and the way of resistance growth depends on the rate of energy input, determined by the current density j.

This criterion divides explosions into several types [4], but these can be broadly grouped into "fast" and "slow" on the basis of a comparative process uniformity along the length of the wire [9]. Data [9] on exposure of the wire in the water clearly show the boundary between these types of electrical explosion. It is within the range the considered fuse. In this connection, it is expedient to identify the main factors influencing on the electrical characteristics of the “fast” and “slow” explosion.

Defining process for a quick explosion are relatively uniform along the length of the wire metal boiling and scattering of metal particles. Assuming that the destruction occurs as an evaporation wave moving from the surface wire to its axis [10], it is possible to introduce the typical scattering speed ϑ, which depends on the injected energy. Scattering time is proportional to the diameter of the wire $t_p=d/2\vartheta$, and the time ratio $t_p/t_н$ form fast explosion similarity criteria

$$P = t_p/t_n = j^2\, d/2h\vartheta \quad (3)$$

This dependence the scattering time of the wire diameter is observed for fast enough explosions in the air [6]. It leads to the fact that the maximum voltage is different for conductors with different diameters, but with the same general cross-section, exploded in the same conditions. This fact is the basis for conclusion of electrical explosion similarity criteria laid down in [11].

Similarity criteria containing complex j^2d has been obtained in [12] by a few other prerequisites. At constant characteristics of the wire material and environmental values of $h$and ϑ can be omitted **and** the criterion takes the form j^2d. All the explosion characteristics will be functions of j^2d, in particular, specific resistance *ρ_m* at the time of the maximum voltage

$$\rho_m = U_m/lj = f(j^2d) \quad (4)$$

* Specific values: current density and resistivity are calculated per unit of the initial cross-section of the conductor.

With energy input rate decreasing a decisive role in the process EVP acquire the magnetic field forces, leading to the formation of instabilities along the length of the wire [9, 13]. As a result, there are briges, which release nearly all the injected energy. In place of the briges formed arc, the number of which is increasing over time, and which causes a voltage growth on the conductor [14]. The time constant of the MHD instabilities is back proportional to the current density:

$$\tau_{MHD} = \sqrt{2\gamma/\mu_0/j},$$

where μ_0 is the magnetic permeability of vacuum, γ is the metal density.

In this model the voltage on the conductor should depend on the current density. At explosion in the water short wires, which does not completely off the current in the circuit, there is evidence [15] that the maximum voltage varies by 25% only when the diameter change to 4 times while maintaining the general cross-sections. At the [16] for electrical explosion in condensed media noted dependence $E_m=U_m/l$ on the wire diameter. There is a graph of changes E_m of the current density for copper conductors in the range $j=(10^9\text{-}10^{11})$ A/M^2.

Unfortunately, wide zone E_m (gradient values differ by about 2.5 times) is not possible to solve the problem of calculating the electrical characteristics of an exploding wire in the oil.

3. Experimental Part

Due to such a controversial baseline information on the independent variables and the relationship between them, as well as the insufficient number of published data for engineering calculations for choosing the length and cross sections of exploding wires, the experiments to investigate the characteristics of exploding wires in transformer oil media and water media are conducted.

In processing the results of experiments as the dependent variable is taken maximum apparent resistivity ρ_m, because it is more stable than the maximum a voltage gradient and, in addition, it is possible to specify the scale of it - initial resistivity ρ_0, while for the large-scale gradient values of the parameters that define our phenomenon it can not be created.

Using the obtained data the maximum resistivity dependence $\rho_m=E_m/j$ in functions of j^2d and j, i.e. from the standpoint of "fast" and "slow" explosion, have been constructed (Figure 2).

Preliminary analysis showed that the spread of the points about the regression line in both cases is approximately the same - 20%. Due to the same accuracy of the description, a more simple dependence for the practical use are encouraged - dependence on the current density. Approximating all the experimental point line (Figure 2) can be described by the expression

$$\rho_m = -0.29 \times 10^{17} \times j + 5.25 \times 10^{-6} \quad (4)$$

This dependence is consistent with the results of [16] with the appropriate terms (in the E_m and j coordinates). The experimental results are also determined by the dependence of the specific action explosion h from the current density (Figure 2).

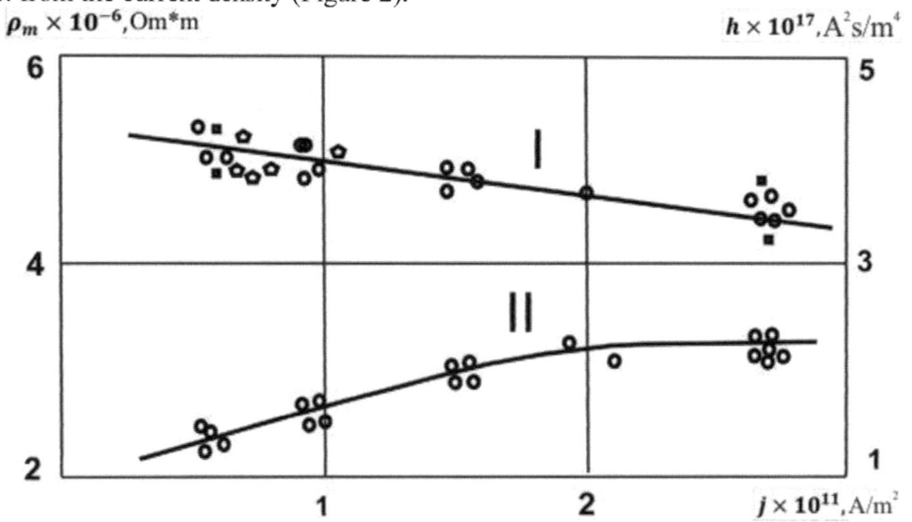

Figure 2. The dependence of the maximum resistivity ρ_m and specific actions h on the current density j, where **O** are experiments in the LC circuit in the transformer oil medium; ◘ are experiments LC circuit in the raw water; ⬟ are experiments on the model of the DRV.

The graph shown in Figure 3, is the basis for the choice of the cross-section of an exploding wire at known its length. In this case, it is required to determine the overvoltage value and the moment of its appearance after the current switched to the second stage. The amplitude of the voltage peak is determined by the approximating expression (5).

$$U_m = l(-0.29 \times 10^{17} j^2 + 5.25 \times 10^{-6} j) \quad (5)$$

Heating time is determined from the equation (1). However, the value depends on the current density. Therefore, it is expedient to use a graphical method of finding the optimal cross-sections of the wire.

Figure 3 shows the electrical characteristics match of exploding fuse wires and in the LC-circuit and second stage of two-stage explosive type switches (in transformer oil and water media). Latter allows to identify the impact on the voltage impulse of the two media and transfer obtained data to other applications of exploding wires.

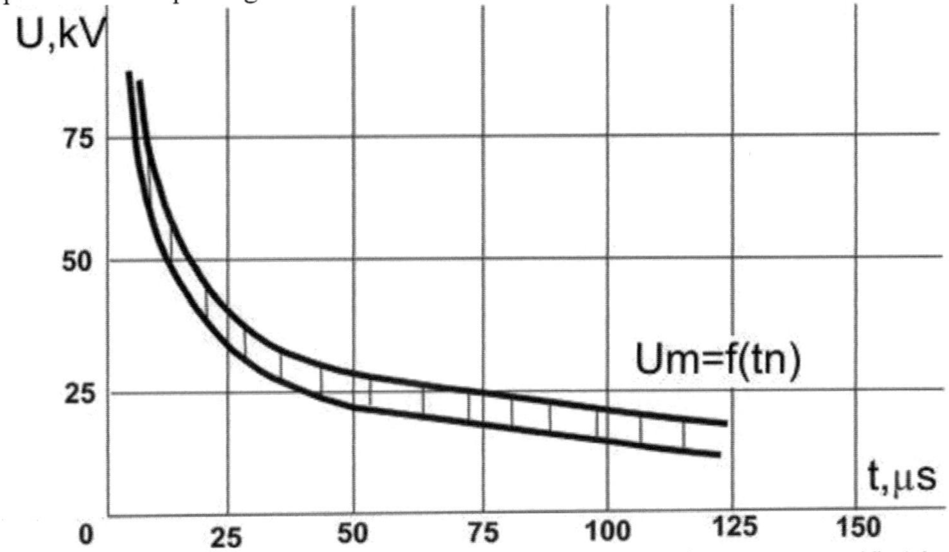

Figure 3. Overvoltage dependence on the wire heating time. Overvoltage range specified for the wire length of 100 mm.

4. Conclusion

The experiments conducted to investigate the characteristics of exploding wires in transformer oil media and water media made it possible to conclude about the typical scattering speed, which depends on the injected energy, scattering time and the time ratio form fast explosion similarity criteria.

Using the obtained data the maximum resistivity dependence in functions of j^2d and j, i.e. from the standpoint of "fast" and "slow" explosion, have been constructed. It made possible to identify the impact on the voltage impulse of the two media and transfer obtained data to other applications of exploding wires.

The obtained results help to reduce the tripping time as much as possible, and limit the amplitude of the fault current, that is very important for saving of working capacity of life-support systems.

References

[1] Babinovich D E *and* Shutov E A 2014 *WIT Transactions on Ecology and the Environment* **190** 501–510 DOI: 10.2495/EQ140481
[2] Lukutin B V *et al* 2016 *IOP Conf. Ser.: Mater. Sci. Eng.* **127(1)** 012005 DOI: 10.1088/1757-899X/127/1/012005
[3] Lukutin B V *et al* 2014 *Proc. of the Int. Conf.* EEEIC'14 459-463 DOI: 10.1109/EEEIC.2014.6835913
[4] Obuhov S *et al* 2016 *MATEC Web of Conferences* **91** 01036 DOI: 10.1051/matecconf/20179101036
[5] Korol'kov V. L. *and* Sivkov A. A. 1983 *Elektrotekhnika* **4** 21–23 (in Russian)
[6] Rozanov J K *et al* 2001 *Electric and electronic devices* (Moscow: Informelectro) (in Russian)
[7] Rukhadze A A (Ed) 1965 *Electric explosion of conductors* (Moscow: Mir) (in Russian)
[8] Burtsev V A *et al* 1977 *Electric explosion of conductors* (Moscow: NIIEFA) (in Russian)
[9] Litvinenko V P 1983 *Physical basis of electric explosion* (Kiev: Naukova Dumka) (in Russian)

IOP Conf. Series: Materials Science and Engineering **189** (2017) 012004 doi:10.1088/1757-899X/189/1/012004

[10] Bennett F D 1974 *Evaporation wave* (Moscow: Mir) (in Russian)
[11] Azarkevich E I *et al* 1975 *Sov. Phys. Tech. Phys.* **20(1)** 111–112
[12] Ivanov V V 1983 *TVT* **21(1)** 146–154
[13] Abramova K B *et al* 1966 *DAN SSSR* **67(4)** 778–781 (in Russian)
[14] Bakulin Y D *et al* 1976 *Sov. Phys. Tech. Phys.* **21(9)** 1144–1151
[15] Krivitskiy E V and Litvinenko V P On the mechanism of explosion of conductors by current pulses. - Technical Physics, 1976, t.46, №10, p. 2081-2087
[16] Krivitskiy E V *and* Litvinenko V P 1976 *Sov. Phys. Tech. Phys.* **46(10)** 2081–2087
[17] Budovich V L *and* Kuzhekin I P 1975 *Proc. of the MEI* **224**
[18] Korol'kov V L *et al* 1986 *Electrical* **9** (in Russian)

IOP Conf. Series: Materials Science and Engineering **189** (2017) 012005 doi:10.1088/1757-899X/189/1/012005

Day-Ahead Short-Term Forecasting Electricity Load via Approximation

R N Khamitov[1], A S Gritsay[2], D A Tyunkov[3], and G E Sinitsin[4]

[1]Professor, Omsk State Technical University, Omsk, Russia
[2]Senior Lecturer, Omsk State Technical University, Omsk, Russia
[3]Assistent, Omsk State Technical University, Omsk, Russia
[4]Head of Electricity markets department, LLC "Omsk Energy Retail Company", Omsk, Russia

E-mail: aleksandr.gritsay@gmail.com

Abstract. The method of short-term forecasting of a power consumption which can be applied to short-term forecasting of power consumption is offered. The offered model is based on sinusoidal function for the description of day and night cycles of power consumption. Function coefficients – the period and amplitude are set up is adaptive, considering dynamics of power consumption with use of an artificial neural network. The presented results are tested on real retrospective data of power supply company. The offered method can be especially useful if there are no opportunities of collection of interval indications of metering devices of consumers, and the power supply company operates with electrical supply points. The offered method can be used by any power supply company upon purchase of the electric power in the wholesale market. For this purpose, it is necessary to receive coefficients of approximation of sinusoidal function and to have retrospective data on power consumption on an interval not less than one year.

1. Introduction

Relevance of work is caused by development of the wholesale market of the electric power and capacity and need of the companies, working at it to creation of short-term models for implementation of process of forecasting of power consumption for the days ahead. From quality of such models, eventually the financial result of the company depends. Now there is a big arsenal of methods of creation of short-term forecasts, but they aren't always effective for such forecasts as they strongly depend on dynamics of power consumption. Besides, other problem characteristic of power supply companies is a lack of opportunities of collection of hourly indications of metering devices of a large number of consumers. At the same time, power supply companies are, as a rule, limited to data of the actual hourly accounting of a power consumption of clients and operate in case of creation of the forecast with data of the system operator on delivery points. On the one hand, it simplifies process of aggregation of data, on the other hand, complicates process of creation of predictive model. The offered technique allows, operating with actual data of the system operator on delivery points, to construct prognostic model which considers dynamics of power consumption. At the same time, process of power consumption is described by an approximating sine function which coefficients – the period and amplitude are with use of the device of an artificial neural network.

IOP Conf. Series: Materials Science and Engineering **189** (2017) 012005 doi:10.1088/1757-899X/189/1/012005

2. Research

In [1] authors the known methods of short-term forecasting of power consumption are systematized. The greatest distribution was gained by methods of the "structural models" group. First of all, it is connected with a possibility of processing of big data arrays by means of modern computing systems, and also development of intellectual algorithms and methods of formation of the training selections for them. In works [2–3] the methods based on use of the device of an artificial neural network were considered, the neural network in this case is presented to the expert in the form of a black box with inputs and outputs. In [4–6] works based on a method of the principal components, but as these methods belong to multiple-factor models and require a large number of signs for creation of the forecast, feasibility of their use isn't justified in case of absence of data of hourly accounting a large number of consumers. In the real article by authors the method of approximation of the power consumption by function of a sine, function coefficients – the period and amplitude which are selected taking into account character of power consumption, using the device of an artificial neural network is offered. Besides, the residuals of volume of a power consumption received by a difference of the actual and expected values are also predicted for the day ahead with use of the device of an artificial neural network.

If to consider the daily schedule of power consumption, according to the available data of power supply company, presented in the Figure 1a; it is possible to notice that in a form it reminds a sinusoid. Besides, the cost of the electric power can also be described sinusoidal function as in hours of the maximum power consumption and cost on the electric power will be higher. Considering power consumption in monthly scale, it is possible to notice that in this case the schedule also reminds sinusoidal function, but already with the week period. In this connection, authors have made a hypothesis of a possibility of implementation of the short-term forecast by an approximation method sinusoidal function. For extension of the approximating function for the days ahead it is required to receive her coefficients by means of which in the subsequent, it could be prolonged for an interval till 24 o'clock.

As follows from the schedule of power consumption that day and night cycles have obvious distinctions on character of power consumption, it is offered to use two sinusoidal functions which would describe separately day and night cycles of power consumption.

3. Experiment

The general formula of the sinusoidal function shows in Equation 1:

$$y = A\sin(kx + b) + D, \qquad (1)$$

where y is the calculated electricity load, MW; x is an hour of the day; k is frequency; b is phase; $A = \dfrac{\max(PC) - \min(PC)}{2}$ is amplitude; $D = \dfrac{\max(PC) + \min(PC)}{2}$ offset concerning a power consumption axis; max(PC) and min(PC) is the maximum and minimum power consumption on the selected time interval respectively. For determination of coefficients of k and b the least-squares method [7] was used. The received coefficients are provided in Table 1.

Table 1. Calculated values of parameters k and b.

Coefficient	Night cycle	Day cycle
k	0.4	0.44
b	4.4	0.8

IOP Conf. Series: Materials Science and Engineering **189** (2017) 012005 doi:10.1088/1757-899X/189/1/012005

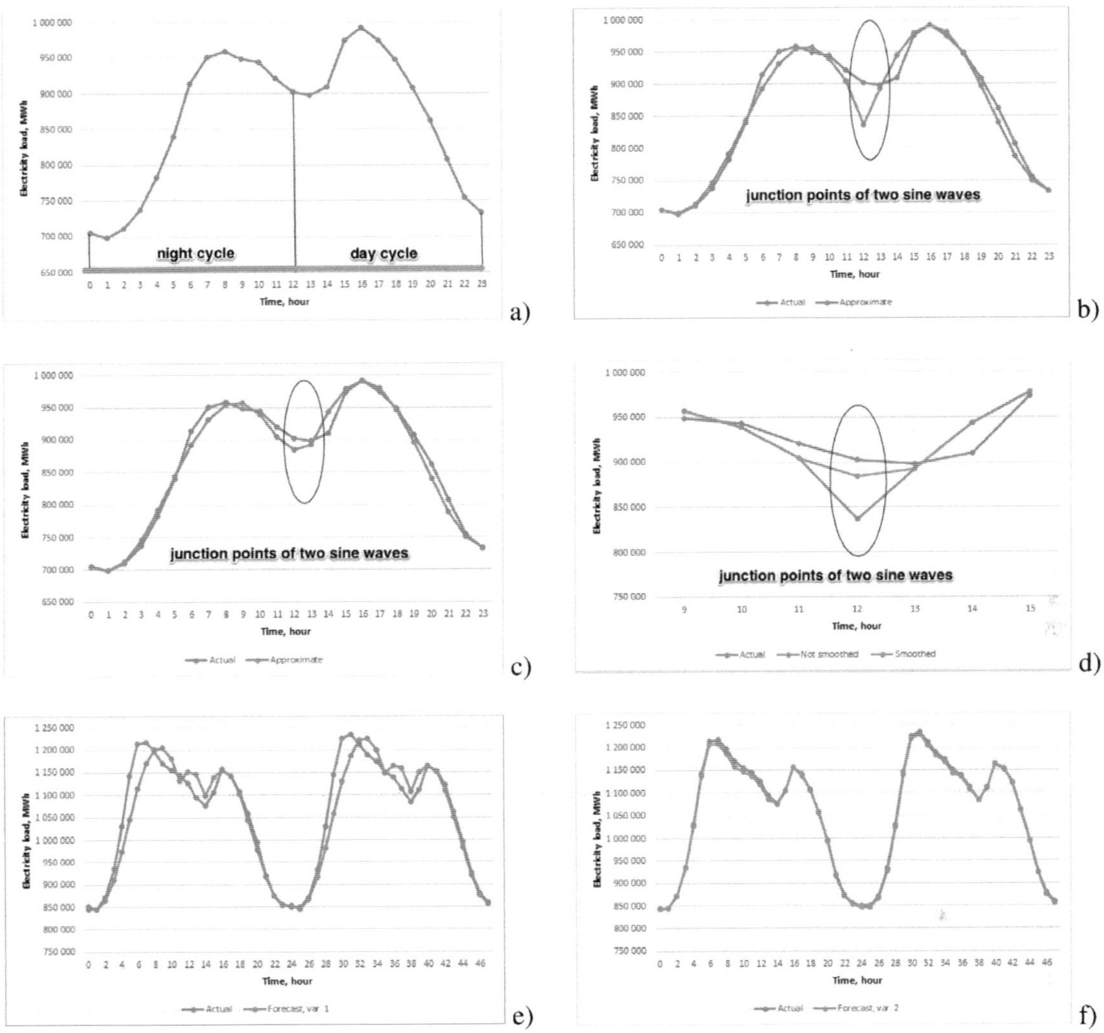

Figure 1. Available data of power supply company, where a is the graph of power consumption on an interval a daily interval (24 hours); b is comparing of graph of the actual and estimated power consumption on a daily interval; c is comparing of graphs of the actual and smoothed by a polynomial of the third level estimated power consumption on a daily interval; d is comparison of actual and smoothed by a polynomial of the third degree of connection points of sinusoidal waves day and night cycles; e is actual and forecast data, the method of approximation for 24 hours; f is actual and forecast data by approximation and forecast of residues obtained by 24 hours in advance.

Figure 1b shows actual and approximated sinusoidal function separately for the graphs of day and night cycles of power consumption. From a Figure 1b, it is visible that the greatest discrepancy of the actual and estimated values is watched in a point of connection of sinusoids which describe day and night cycles of power consumption. At that time, the average error of approximation on an annual interval made 4.52% (real data on a power consumption for 2015 of power supply company were used), in a point of connection of the sinusoids describing day and night cycles of a power consumption, average deviation on an annual interval made 14.52% that considerably exceeds an average error on other hour intervals. In order of reduction of value of an error of power consumption in these point different methods of approximation were tested. The received results, when using these methods, are provided in Table 2.

IOP Conf. Series: Materials Science and Engineering **189** (2017) 012005 doi:10.1088/1757-899X/189/1/012005

Table 2. Comparing of methods of approximation for finding of a point of connection of sinusoids for day and night cycles.

No.	Approximation method	The reliability on 1 year data interval	The reliability on 24 hours' data interval
1	Linear	0.01	0.13
2	Logarithmic	0.01	0.20
3	Exponential	0.01	0.24
4	Polynomial of degree 2	0.15	0.84
5	Polynomial of degree 3	0.15	0.92

The best result of approximation of points of connection of sinusoids by a polynomial of the third level was achieved by using two next values to the calculated points and starting and finite value of selection. For finding of values of unknown members of model it is possible to use any known method [8]. The received result of approximation is presented in a Figure 1c, peak discrepancies of the actual and estimated values of power consumption in points of connection of the sinusoids describing night and day cycles were removed. In a Figure 1d the graph of the actual power consumption and received by an approximation method before smoothing by a polynomial of the third level, and after smoothing is presented.

For implementation of the short-term forecast of a power consumption the offered method of approximation requires to find coefficients of A and D on a step of $X+1$ where X is the current days, used a neural network [9]. The neural network consists of three full-meshed layers of neurons where all neurons of one layer are connected to neurons of the following layer. Function of activation of neurons is sigmoid. For training of a neural network Levenberga-Markvardt's algorithm was used [10]. For training of a neural network the learning selection consisting from six input and one output parameter was created. In Table 3 the selection fragment for prediction for 02/17/2016 of night and day cycles of power consumption is shown. Mathematically dependence of coefficient of A can be shown function from variables (Equation 2):

$$A_{res} = \{A, C, V_{avg}, T_{avg}, h\}, \tag{2}$$

where C is power consumption cycle (0 – night, 1 – day); T_{avg} is average daily ambient air temperature (this parameter is the major, consuming of electrical energy directly depends on it), °C; h is the sign specifying whether day is holiday or working (in case of h=1 – the day off, h=0 – the working day); t is an hour interval of time; V_{avg} is the consuming approximated by a sinusoid, A is the actual value of coefficient of A, A_{res} is expected value of coefficient of A.

Table 3. Fragment of training sample for A forecasting coefficient on 02/17/2016.

Input neural network						Output neural network	
Date	A	C	V_{avg}	T_{avg}	h	A_{res} night	A_{res} day
02/10/16	144062 160350.5	0 1	921482.5 935633.5	-10.9	0		
02/11/16	141897.5 153709.5	0 1	909551.3 917336.3	-8.1	0	135411.6	172066.8
02/12/16	91956 143566	0 1	820419.75 868883.4	-5.3	0		

The coefficient of D was found similarly, at the same time on an input of a neuronet the coefficient of D and as the expected D_{res} value moved. The result of approximation of a function graph by method of finding of coefficient t with use of a neural network is shown in a Figure 1e. From the diagram, it is visible that expected values are offset rather actual diagram in this connection, the decision to realize prediction of residuals of volume of power consumption, values, received method of a difference actual and expected with use of a sine function, was made. Forecast was also carried out with use of the device of an artificial neural network. Learning selection for the forecast of residuals for a day sinusoid is presented in Table 4.

Table 4. Fragment of training sample for forecasting residues on 02/17/2016.

Input neural network							Output neural network
Date	d	t	V	$V_{\sin}$	T_{avg}	h	d_{res}
02/10/16	-2465.3	0	748703	751168.3	-10.9	0	-5256.6
	-552.5	1	744196	744748.5			-637
	-474.8	2	760511	760985.8			2800
	28913.5	3	826230	797316.5			28723
	75398.2	4	923403	848004.8			68214.1
	93778.7	5	998827	905048.3			102262.8

In Table 4 the fragment of learning data, for forecast of the residuals of volume of a power consumption received by method of a difference actual and expected (an approximation method) of power consumption volumes is shown, where d_{res} is expected value of residual, [MW]; d is actual value of residual, [MW]; t is an hour interval of time; $V_{\sin}$ is value of the power consumption received by method of approximation of a sine function, [MW]; V is the actual value of a power consumption, [MW].

The result is presented in Figure 1f. As it follows from the Figure 1f, the method of function approximation and residues forecasting with the use of artificial neural network brought the forecast values and the actual values together, so it possible to use this approach in compiling forecasts for Wholesale Electricity and Capacity Market (WECM).

4. Result

The process of power consumption short-term forecasting by approximation in the form of a UML schema is shown in Figure 2. The power consumption process is known to be of different character on different days [11]. The realized process takes into account the specifics of forming a variety of training samples for retrospective data: filter 1 generates a data set for training sample of week days, filter 2 is for weekends and filter 3 is for holidays.

The algorithm for power consumption short-term forecasting was realized using Rapidminer analytical system. It is distributed under the LGPL (Lesser General Public License) and has a flexible customizable structure with many realized operators. If it is necessary to implement the algorithm not being available in the standard configuration of the analytical information system, it is possible to extend it through the implementation of the interface developers specific system. This is possible using JAVA programming language and integrated development environment Eclipse. The resulting extension is compiled into a library file and it connects Rapidminer platform within startup process.

The fragment process of power consumption approximation using Rapidminer analysis system is shown in Figure 3.

5. Conclusion

The offered method approximation by a sine function, coefficients – amplitude and the period which are set up with use of an artificial neural network and the residuals of volume of a power consumption received by a difference of the actual and expected values, also predicted with use of an artificial neural network allowed, based on retrospective data on a power consumption separately to the taken power supply company, to provide the accuracy of 1.5% on an annual interval of data. This method can be used by the subject of the wholesale market of the electric power, not only power supply company, but also the industrial enterprise. For this purpose, it is necessary to enumerate coefficients of *k, b, A, D* sine function of approximation by the offered method. The algorithm can be realized with use of analytical system, or is presented in the program form.

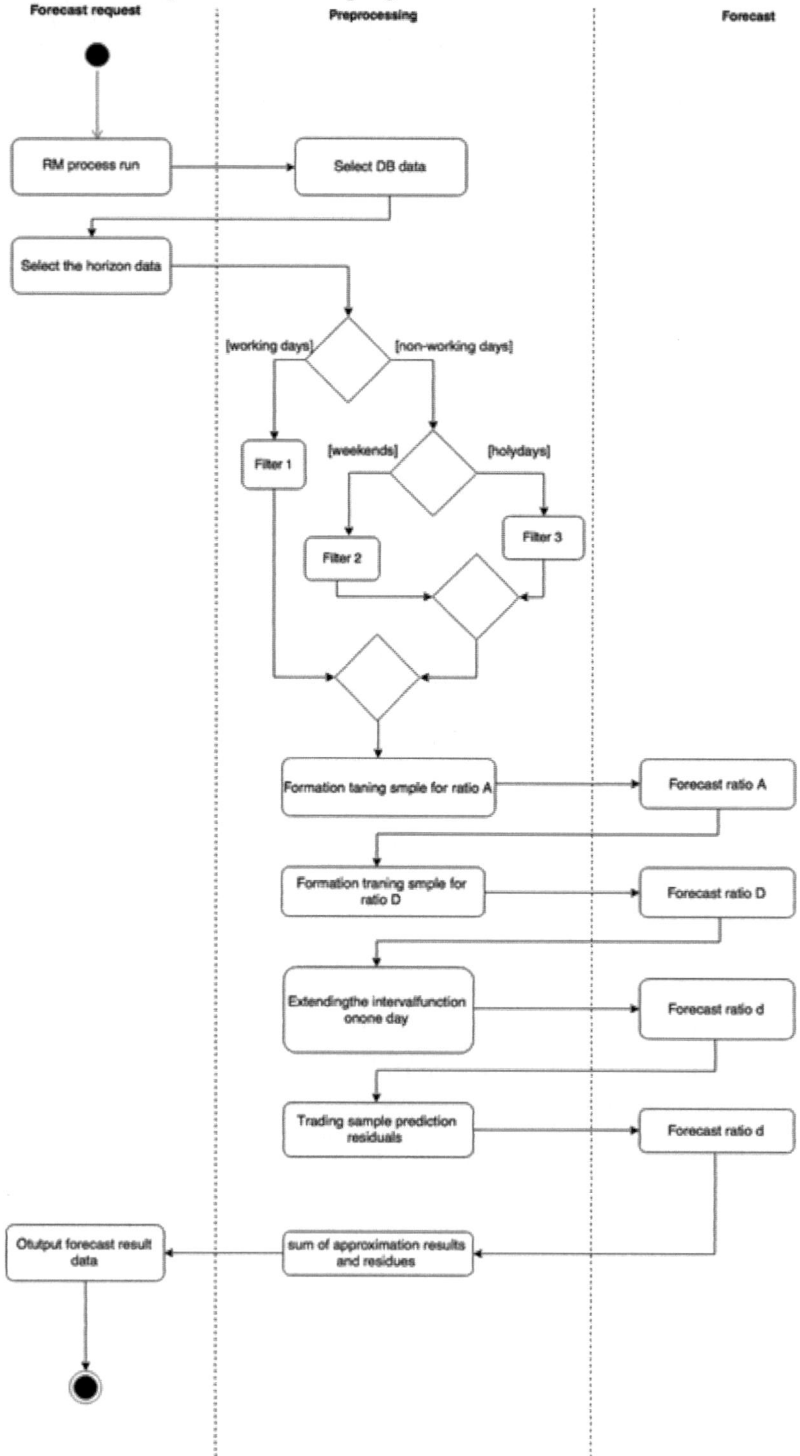

Figure 2. UML-diagram of electricity load short-term forecasting by approximation method.

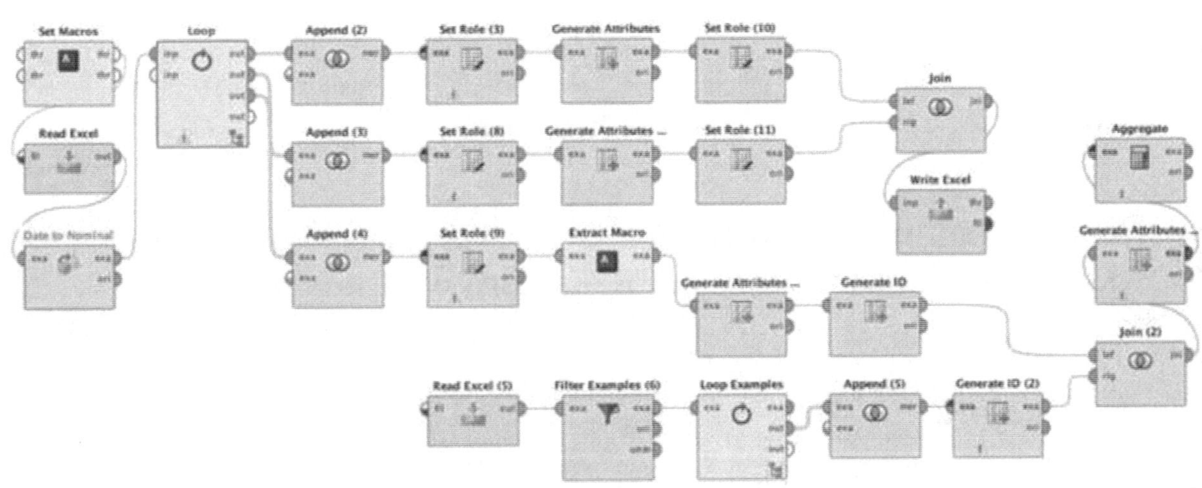

Figure 3. Fragment algorithm of approximation of electricity load implemented in analytical system Rapidminer.

References

[1] Gritsay A S *and* Tyunkov D A 2016 *Proc. of All-Russian Conf. Actual Energy Issues* **1** 41–45 (in Russian)

[2] Han P *et al* 2010 *Mathematical and Computer Modelling* **51(11–12)** 1398–1403 DOI: 10.1016/j.mcm.2009.10.031

[3] Panapakidis I P *and* Dagoumas A S 2016 *Applied Energy* **172** 132–151 DOI: 10.1016/j.apenergy.2016.03.089

[4] Gao C *et al* 2007 *Physica A: Statistical Mechanics and its Applications* **382(1)** 98–113 DOI: 10.1016/j.physa.2007.03.050

[5] Borikov V N *et al* 2016 *J. Phys.: Conf. Series* **671(1)** 012014 DOI: 10.1088/1742-6596/671/1/012014

[6] Nadtoka I I *and* Balasim M A 2015 *Procedia Engineering* **129** 657–663 DOI: 10.1016/j.proeng.2015.12.087

[7] Arbind A *and* Reddy J N 2016 *Int. J. of Non-Linear Mechanics* **79** 38–47 DOI: 10.1016/j.ijnonlinmec.2015.10.009

[8] Bini D *et al* 2016 *Linear Algebra and its Applications* **502** 1–4 DOI: 10.1016/j.laa.2016.03.042

[9] Potapov V I *et al* 2016 *Bull. of the Tomsk Polytechnic University Geo Assets Engineering* **327(8)** 44–51 (in Russian)

[10] Amini K *and* Rostami F 2016 *J. of Computational and Applied Mathematics* **300** 30–42 DOI: 10.1016/j.cam.2015.12.013

[11] Shcherbakov *et al* 2010 *Bull. of the Volgograd State Technical University* **9** 68–71 (in Russian)

IOP Conf. Series: Materials Science and Engineering **189** (2017) 012006 doi:10.1088/1757-899X/189/1/012006

Electromagnetic Method for Exogenetic Geodynamic Elements Mapping in Permafrost Environment

V F Gordeev[1], S Y Malyshkov[1], and V I Polyvach[2]

[1]Senior Researcher, Institute of Monitoring of Climate and Ecosystems, Siberian Branch of the Russian Academy of Science, Tomsk, Russia
[2]Junior Researcher, Institute of Monitoring of Climate and Ecosystems, Siberian Branch of the Russian Academy of Science, Tomsk, Russia

E-mail: msergey@imces.ru

Abstract. Taking into account the global warming, there is a pressing need to detect thermokarsts and to monitor permafrosts during the design and construction of industry infrastructure in Northern regions. The paper suggests a permafrost probing method based on the Earth's natural pulsed electromagnetic field parameters recording. Authors describe the architecture and algorithms for the recording hardware. Examples of thermokarst detecting in poleward region are demonstrated. Pulsed electromagnetic fields intensity over thermokarst funnels is 30 times higher than the background levels. Authors substantiate the method's eligibility to monitor geocryologial processes.

1. Introduction

Natural climate dynamics and technogenous changes to surface conditions due to design and maintenance of engineering structures cause a pressing need to estimate a soil bearing capacity and monitor soil conditions. Over 45% of territory in Russia is prone to thermokarst. 72 cities are impacted by it. Thermokarst problem draws considerable attention lately due to global warming and hazardous geocryological processes intensification. Further development of those trends can lead to considerable terrain variations at many industrial sites with rather negative consequences. Rules and regulations for engineering and geological survey for construction at perennially frozen ground determine requirements for spatiotemporal forecast of the surveyed terrain engineering-geocryological conditions to retain constructions stability. Calculated forecast time frame is generally determined by the qualified lifetime of facilities designed, as changes will happen and influence the facilities throughout their whole lifetime. Changes originating from natural causes usually manifest themselves rather slowly, but there are short term unpredictable climate fluctuations and technogenous changes to geocryological conditions which are impossible to account for in a long term forecast, which is confirmed by a large share of deformed permafrost built constructions. Therefore it is necessary to observe ground conditions dynamics, especially at potentially hazardous industrial sites.

2. Results

Earth's natural pulsed electromagnetic field (ENPEMF) method allows to locate all the fault lines in rock masses in place, including fractures, raptures, sinkholes, taliks, other inconsistencies and to monitor their dynamics to predict potential industrial site hazard. Radio-wave methods applied are based on the phenomenon of electromagnetic emission - dielectric materials emissive ability when they are acted on. Electromagnetic emission emerges in the process of charges generation and relaxation on fracture planes during the stress state of the rocks. Pulses emerge both when dielectric uniformity changes and when electrolyte-filled capillars rift. In rock formations there can be the following sources of natural electromagnetic fields: soil structure inconsistencies, unequally strained structures, fractures and microfractures. All those sources generate pulsed electromagnetic fields as a result of mechanic-to-electric energy conversion, strain waves from the mantle, tides, microseisms, winds, and technological loads thus creating a natural electromagnetic background of lithospheric nature. Observing electromagnetic emission allows to monitor stress-strained state of the rock formation. In a variable field of tectonic stresses (stress-strain) rocks, containing dielectric minerals emit electromagnetic pulses due to mechanical-electric energy conversion. Intensity, frequency and amplitude of those pulses are determined by the structural, strength and deformational parameters of rock masses. In a clay rock, aside from fracturing and deformation of rock matrix, double dielectric layers also take part in electromagnetic fields pulses generation. Intensive fluid filtration in interstices and fractures is also accompanied by the polarization and ENPEMF emergence. Tectonisms attributed to block faulted zones, even slowest and low in amplitude are always accompanied by the structural bonds disturbance and the emergence of micro and macro fractures in rock solids, as well as by the diffuse layer ion density change in rock solutions. Those processes lead to electromagnetic emission. Micro and macro fracturing is attended by the charges generation at the fractures shoulders due to electro-adhesion, fracturing electrification. Those charges generate high intensity electric field inside which there's a possibility of a breakdown in gaseous rock phase, filling microfractures.

Perennial measurement in various regions demonstrated that lithospheric ENPEMF has distinguishable daily and seasonal variations. That can be explained by the fact that stress-related waves in Earth's crust are related to Earth's nonuniform rotation around own axis and around Sun [2]. Daily variations depend on calendar date, geolocation, location's geophysical attributes [3]. Pulsed electromagnetic fields can change both with ground conditions changes and with field sources influence changes. For example, typical daily variations can be disrupted by the changes in the Earth's crust rhythmic movement as a result of separate crust blocks joining into consolidated mass prior to earthquake or when stress-strained state of rock changes. So, ENPEMF recording can be a multipurpose tool for geophysical survey, Earth's crust geodynamic activity monitoring and Earth sciences research.

Proof of feasibility test for exogenetic geodynamic elements mapping in permafrost environment was performed at Chayandinskoe oil and gas condensate deposit in Yakutia. Permafrost area at the research location is over 60%. Maximum frozen zone depth on flat and table land is 100-250 m. Maximum depth of seasonal thawing is 3.0-4.6 m for sands and 2.0-2.7 m for clay loam. Thermokarstic formations predominantly develop in very icy alluvial and boggy deposits. Thermokarst processes develop in paragenesis with heaving. The reasons for thermokarst development are increase in yearly average rocks temperature and changes in a terrain rate of water-encroachment. Currently thermokarst processes are developed on table lands of watershed, formed as eluvial deposits on terrigenous-carbonate rocks formations and manifest at small basins of up to 0.5 m in depth. Thermokarst processes are connected to segregated ice thawing. Their development intensified strongly since oil and gas deposit exploration and development.

During the office studies ENPEMF anomalies were compared against engineering-geocryological drilling, snow survey, field and laboratory research of water-physical property of ground and remote morphotectonic analysis library materials. All the date mentioned above was provided by "MSU-Geophysics" Ltd.

ENPEMF recording was performed by the multichannel geophysical recorders "MGR-01" [4]. In basic configuration MGR-01 has two channels measuring magnetic components and on channel measuring electric component of electromagnetic field. Antennas to receive ENPEMF magnetic component in north-south direction and west-east direction are magnetic ferrite antennas, receiving a signal in a very low frequency (VLF) band. Mutually orthogonal position of those antennas allows to obtain system directional pattern close to radial. Differential capacity transmitter, working in near-field reception zone in a frequency band of 500 Hz – 100 kHz was used as an antenna for ENPEMF electric component reception. Amplification path parameters of recorders and settings are changed by software. Recorders allow to measure pulsed flow intensity with a preset sampling interval from 1 second up to 4 minutes and an amplitude of the first pulse exceeding preset threshold for each sampling interval.

Before the start of the work all recorders were set up for the same sensitivity, amplification paths parameters were picked based on geological and tectonic particular properties at the operating region according to calibration charts [3]. Technically ENPEMF spatial anomalies detection was performed by comparing pulsed flow intensity data recorded at each survey station against the reference station data using authentic patented technique. That way background noises are excluded from EMPEMF spatiotemporal variations and anomalies related to Earth's crust structure at measurement site are isolated.

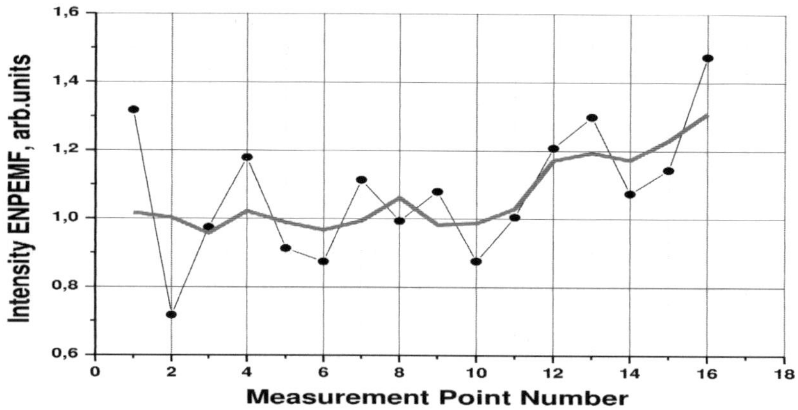

Figure 1. Spatial variations of electromagnetic noises at the locations without rock discontinuity.

Figure 1 represents ENPEMF spatial variations at a location without tectonic disturbance and Earth's crust exogenous geodynamics. At that and other charts thin line with dots shows actual measurement and thick curve shows 3 points running average. Although in general the territory surveyed has a complicated tectonic structure, electromagnetic noise anomalies at this plot do not exceed 40% over the background noises.

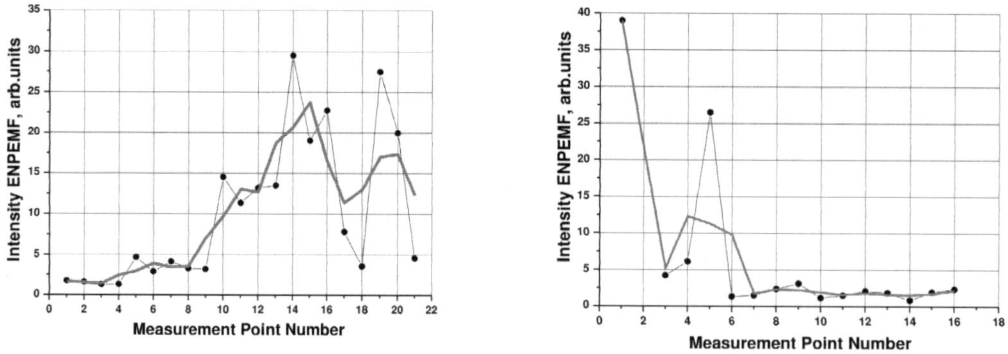

Figure 2. Spatial variations of electromagnetic noises at the locations nearby exogenous geodynamic elements.
a) thermokarst funnel; b) sinkhole.

Figure 2 represents ENPEMF spatial variations when crossing thermokarst funnels. On those charts anomalous noise values are 30-40 times higher than background values. Based on the integrated geophysical research of the location nearby to specified Earth's crust disturbance is assigned to very dangerous category by engineering-geocryological conditions. Complexity of development conditions category is very complex. Perennially frozen grounds have sporadic distribution.

ENPEMF method applicability for stress-strained state of rocks dynamic monitoring is confirmed by successfully operating systems for geohazard prediction at main gas pipelines [3-5]. ENPEMF parameters based algorithm for geodynamic processed monitoring is based on two criteria. First one shows difference in reading intensity between the recorders located at stress-strained state of rocks observing points and the reference recorder. Reference recorder is located not farther than 25m (VLF band wavelength) from observing ones at a site safe from geodynamic activity. Another criterion assesses similarity in different recorders data. Excess intensity of ENPEMF pulse flow at each observation point comparing to reference station is calculated based on specific formulas.

As a second parameter of stress-strained state of rocks, estimation is used for correlation between ENPEMF daily variation of reference station and i-th observation station. Spearman's formula [6] is used to calculate the correlation coefficient.

Pulse flow intensity surplus of over 150% over three days and more is considered a potential hazard for industrial site operation.

3. Summary

Suggested method for ENPEMF signal analysis and a network of spatially distributed stations provides accurate reproducible results, robustly reflecting geodynamic activity. System application allows to detect tension and compression zones, stress directions in real time for rock structures with landslide activities. A system equivalent to the one described above could be used to monitor perennial frozen ground state [7].

In summary, Earth natural pulsed electromagnetic fields method can be applied for both preliminary surveys at permafrost engineering sites and monitoring of permafrost under technogenous influence.

References

[1] Malyshkov Yu P *and* Malyshkov S Yu 2012 *Eccentric rotation of the earth's core and lithosphere: Origin of deformation waves and their practical application* (Taiwan: Nova Science Publishers) **6** 115–212

[2] Malyshkov Yu P *and* Malyshkov S Yu 2016 *IOP Conf. Ser.: Earth Environ. Sci.* **48** 012028 DOI: 10.1088/1755-1315/48/1/012028

[3] Malyshkov Yu P *et al* 2013 *Electromagnetic Fields: Principles, Engineering Applications and Biophysical Effects* (Taiwan: Nova Science Publishers) 1–86

[4] Shtalin S G *et al* 2012 *Sensors & Systems* **4** 32–37 (in Russian)

[5] Gordeev V F *et al* 2016 *IOP Conf. Ser.: Earth Environ. Sci.* **48** 012026 DOI: 10.1088/1755-1315/48/1/012026

[6] Myers J L *and* Well A D 2003 *Research Design and Statistical Analysis* (Mahwah, New Jersey & London: Lawrence Erlbaum Associates Publishers)

[7] Malyshkov S Yu *et al* 2016 *Proc. of Int. Conf. GEO-SIBIR'16* **2** 13–18 (in Russian)

IOP Conf. Series: Materials Science and Engineering **189** (2017) 012007 doi:10.1088/1757-899X/189/1/012007

Synthetic Reference Materials Based on Polymer Films for the Control of Welding Fumes Composition

O V Kuznetsova[1], A N Kuznetsova[2], and L A Begunova[1]

[1]Associate Professor, Irkutsk National Research Technical University, Irkutsk, Russia
[2]Student, Irkutsk National Research Technical University, Irkutsk, Russia

E-mail: olvlku20@mail.ru

Abstract. Analysis of the current hygienic situation in the welding production showed that the intensification of welding processes involves the deterioration of air quality, which negatively affects the welders health. Welders are exposed to a variety of metal fumes, including manganese that may elevate the risk for neurological diseases. The control of metals concentration in the air of the working area is difficult due to the lack of reference materials. The creation of reference materials of welding fumes composition is a challenge due to chemical characteristics of their physical properties. Synthetic samples in a form of the polymer film containing powder particles of welding fumes were create. Studies on the selection of the polymer were done. Experiments proved that the qualitative materials of synthetic welding fumes are obtained by using polyvinyl alcohol. The metals concentration in the samples was determined by X-ray fluorescence analysis. The obtained data demonstrates indirectly the uniform distribution of welding fumes powder particles on the polymer film.

1. Introduction

Welding as a process of high-performance production of permanent joints is widely used in the manufacture of metallurgical, chemical and energetic equipment, the installation of various piping, mechanical engineering, shipbuilding, manufacturing of building and other structures. In spite of the achievements in the development of high-performance equipment and the introduction of automated and robotic lines, the proportion of welding performed manually is not reduced, and as a result, the welder is exposed to welding fumes. Elevated concentrations of fumes in the air of a working area can cause deviations in a health status and at a certain duration and intensity of exposure it can lead to the development of occupational diseases, such as bronchitis, respiratory tract irritation, fibrosis, pulmonary cancer [1]. The greatest risk for the occurrence of a respiratory disease is a solid component of welding fumes (SCWF), toxic compounds containing metals such as manganese, iron, chromium, nickel [1, 2]. Therefore, the problem of qualitative laboratory control for the condition of air in the working area is very relevant, and its role is increasing with the wear and tear of processing facilities and with the introduction of modern technologies.

The introduction of rapid methods of SCWF analysis is limited to the absence of national reference materials (RM) of welding fumes composition collected on a filter. There are only a few copies of reference materials found in the foreign analytic practice, for example, BCR-545 (IRMM, Belgium), HSL MSWF–1 and HSL SSWF 1 (UK). BCR–545 material [3–4] is a glass fiber filter loaded with weld dust; it is certified only on the content of total Cr and Cr (VI). HSL MSWF–1 [5] and HSL SSWF–1 materials [6] are welding fumes powder particles collected during welding of soft steel and stainless steel respectively, packed in glass bottles 1 g each; they are standardized on the content of Fe, Mn and Zn.

The specific complexity when creating RM of welding fumes is a wide variation in particle size and the complexity of phase and chemical composition of SCWF. Solid particles of welding fumes have mostly sizes of 10^{-1} to 1 micron regardless of the welding technique [7]. The phase and chemical composition of SCWF depends on conditions of a welding work conduction; therefore, the creation of reference materials for each welding type separately is required.

The certification of RM as real materials collected on the filter is not possible, as it is required to produce a large number of samples copies with the same composition, in order to use a part of them for certification analyses and to use the remaining ones as certified RM. However, uneven distribution of pollutants in the controlled object does not provide a sample group of similar chemical composition. Thus, the problem solution is seen only through the creation of synthetic reference materials, which would be adequate to real materials by their physical and chemical characteristics.

Filters, on which solutions of analyzed elements applied, are used mostly as parts of the analysis of welding fumes as synthetic materials of SCWF composition. For example, in determining the Cr (VI) [8], there were materials used, obtained by applying $K_2Cr_2O_7$ solution on the filter. It should be noted that the filter material may contain varying amounts of different metals [9], so that for samples preparation it is expedient to use the same type and commercial batch of filters as for the sampling. The advantage of this method is the simplicity of preparation, but the main disadvantage is the inadequacy of such SCWF samples to real samples by physico-chemical properties. Therefore, it is more promising to use materials prepared by precipitation of compounds particles of controlled components on the filter. However, such materials are unsuitable for long-term storage and transportation, as well as during the operation in a vacuum, because of the possible crumbling of particles.

The samples, prepared using a specially designed air samplers or fumes generators, are closer to real SCWF samples, loaded onto aspiration filters. In the works [10–11] etc. to manufacture the materials of SCWF composition, the parallel selection of a large number of welding fumes materials was carried out, then part of the materials was analyzed by different methods and there was an installation of contents parameters settings of the filter components. The remaining part of samples was used as RM. The authors [10] have found that there are differences in weight of fumes on the filter and hence the content of elements in the sample, so there was an additional analysis of loaded filters done using an X-ray fluorescence spectrometer with the wave dispersion to eliminate different copies. In works [12–14] for the control of metal content in the working area when welding steel structures using fumes generator, synthetic samples of generated fume from the metal wire using a robotic system was prepared. It should be noted that these techniques make it difficult to prepare a large number of identical materials with different variations of the chemical composition, specific to SCWF samples.

There is a technology for manufacturing synthetic RM of atmospheric aerosols composition in the form of methylcellulose polymer film containing a finely divided powder carrier of identified components [14]. As a carrier of aerosols particles, there is a mixture of RM of soils composition and metal oxides. The developed technology allows preparing the required number of identical samples with wide chemical composition variations and mass of aerosols particles. The closest in physical and chemical properties to real fumes samples collected on the aspiration filter, are the samples in the form of polymer films containing powder – the carrier of identified elements, which is a homogenized finely dispersed SCWF material. The objective of this study is to develop manufacturing techniques of RM of welding fumes composition in the form of a polymer film containing SCWF particles.

2. Experimental

At the stage of polymer selection the use of the following polymers was evaluated: methylcellulose (MC), polyvinyl chloride (PVC), polyvinyl alcohol (PVA), polystyrene (PS), polyvinylbutyral (PVB), copolymer vinylglycide ester with vinylchloride (VGE-VC) after their dissolving in water, ethanol, acetone, toluene, chloroform, dichloroethane, dimethylformamide. The resulting polymer solution was poured onto a cleaned horizontal glass surface and air-dried. Smooth films were received only from solutions of PVA and MC in water, PS in toluene, PVB and VGE-VC in dimethylformamide; in other cases, the dried polymer films contained crystalline inclusions and could not be separated from the glass surface. The films from VGE-VC solutions in dimethylformamide and PS in toluene were very brittle. In addition, it should be noted that toluene and dimethylformamide are toxic solvents. The films of MC solution in water are hygroscopic, which creates certain difficulties when working with them. The foregoing led to the choice of the polyvinyl alcohol as the polymer to create synthetic RM of welding fumes.

Further studies showed that the optimum ratio of thickness-elasticity are in films produced from the PVA solution in water with polymer concentration of 10% in the solution.

The main difficulty is the introduction of SCWF powder material into the polymer solution. The possibility of the powder introduction into the polymer solution with water suspension was studied. There was the uneven distribution of powder and many conglomerates because of the coagulation of fine SCWF particles in the resulting film. Qualitative films were obtained by the introduction of powder particles into the polymer solution with ethyl alcohol. For this purpose, a sample of a powdered material weighing 50 mg was placed in a beaker, 1 ml of alcohol was added, 50 ml of polymer solution was added to the resulting suspended material and it was mixed thoroughly with a glass rod, poured onto cleaned glass disposed on a horizontal surface, and dried. Individual samples with a diameter corresponding to aspiration filters on which SCWF samples selected, were stamped from the resulting film.

3. Results and considerations

According to the developed technology, the produced set of synthetic materials was based on polyvinyl alcohol to the mass range (mg) of the welding fumes powder 30 (films 1 and 2), 50 (films 3 and 4), 100 (films 5 and 6). Two samples were stamped from each film. The metals concentration in the samples was determined by XRF analysis. Studies were performed on an energy-dispersive X-ray fluorescence spectrometer EDX-8000 (Shimadzu, Japan) based on silicon drift detector with thermoelectrical cooling; an X-ray tube with an Rh-anode (work mode: air cooling, voltage 4-50 kV, current 1-1000 μA). Determination results of Mn, Cr, Zn, rationed by the Fe content in samples are shown in the Table 1.

Table 1. The concentrations of elements in the samples, rationed by the iron content.

sample	Mn	Cr	Zn
1-1	0.0317	0.0106	0.00264
1-2	0.0332	0.0091	0.00302
2-1	0.0321	0.0058	0.00292
2-2	0.0387	0.0060	0.00298
3-1	0.0406	0.0070	0.00280
3-2	0.0409	0.0073	0.00292
4-1	0.0358	0.0076	0.00265
4-2	0.0362	0.0081	0.00268
5-1	0.0397	0.0048	0.00274
5-2	0.0375	0.0056	0.00208
6-1	0.0358	0.0064	0.00239
6-2	0.0365	0.0046	0.00183
Mean value	0.0366 ± 0.0020	0.0068 ± 0.0012	0.00264 ± 0.00023

Errors in the identification of elements were decided by the t-test (t(11, 0.95) = 2.20), they form 5÷18%, depending on the element that meets the requirements for the analysis of welding fumes. It should be noted that the obtained data (see Table 1) demonstrates indirectly the uniform distribution of SCWF powder particles on the polymer film.

4. Summary

Based on the foregoing, a method is provided of making synthetic RM of welding fumes composition that are adequate to real samples collected on a filter. Using SCWF material obtained at various types of welding, allows the creation of reference materials with wide variations of phase and chemical composition.

References

[1] Antonini J M 2003 *Crit. Rev. Toxicol.* **33 (1)** 61–103 DOI: 10.1080/713611032

[2] Korczynski R E 2000 *Applied Occupational and Environmental Hygiene* **15 (12)** 936–945 DOI: 10.1080/104732200750051175

[3] Bordunov S V *and* Galtseva O V 2016 *J. Phys.: Conf. Series* **671(1)** 012009 DOI: 10.1088/1742-6596/671/1/012009

[4] Chauhan S K *et al* 2009 *Environ. Monit. Assess.* **156 (1-4)** 407–418 DOI: 10.1007/s10661-008-0493-1

[5] Butler O 2013 *Reference material HSL MSWF-1. Elements in mild steel welding fume* (United Kingdom: Health and Safety Laboratory)

[6] Butler O *and* Musgrove D 2013 *Reference material HSL SSWF-01. Elements in stainless steel welding fume* (United Kingdom: Health and Safety Laboratory)

[7] Berlinger B 2010 *Analytical and Bioanalytical Chemistry* **399** 1773–1780 DOI: 10.1007/s00216-010-4185-7

[8] Oláh Ĺ *and* Pivoluska J 1985 *J. Radioanal. Nucl. Chem.* **96 (6)** 575–584 DOI: 10.1007/BF02165037

[9] Derg T *and* Royset O 1989 *ISP Information Newsletter* **15 (6)** 337 DOI: 10.1016/0960-1686(93)90411-Q

[10] Stacey P *and* Butler O 2008 *Annals of Occupation Hygiene* **52 (4)** 287–295 DOI: 10.1093/annhyg/men016

[11] Anglov J T B *et al* 1993 *Fresenius' Zeitschrift für Anallytische Chemie* **345 (2-4)** 335–339 DOI: 10.1007/BF00322625

[12] Antonini J M *et al* 2006 *Journal of Occupational and Environmental Hygiene* **3 (4)** 194–203 DOI: 10.1080/15459620600584352

[13] Cena L G *et al* 2014 *Journal of Occupational and Environmental Hygiene* **11 (12)** 771–780 DOI: 10.1080/15459624.2014.919393

[14] Korzhova E N *et al* 2004 RF Patent 2239170 *Byull. Izobret.* **30** (in Russian)

IOP Conf. Series: Materials Science and Engineering **189** (2017) 012008 doi:10.1088/1757-899X/189/1/012008

Method for Prediction of the Power Output from Photovoltaic Power Plant under Actual Operating Conditions

S G Obukhov[1], I A Plotnikov[2], O A Surzhikova[3], and K D Savkin[4]

[1]Associate Professor, National Research Tomsk Polytechnic University, Tomsk, Russia
[2]Senior Lecturer, National Research Tomsk Polytechnic University, Tomsk, Russia
[3]Associate Professor, Czech Technical University in Prague, Czech Republic
[4]Student, National Research Tomsk Polytechnic University, Tomsk, Russia

E-mail: igorplt@tpu.ru

Abstract. Solar photovoltaic technology is one of the most rapidly growing renewable sources of electricity that has practical application in various fields of human activity due to its high availability, huge potential and environmental compatibility. The original simulation model of the photovoltaic power plant has been developed to simulate and investigate the plant operating modes under actual operating conditions. The proposed model considers the impact of the external climatic factors on the solar panel energy characteristics that improves accuracy in the power output prediction. The data obtained through the photovoltaic power plant operation simulation enable a well-reasoned choice of the required capacity for storage devices and determination of the rational algorithms to control the energy complex.

1. Introduction

The most serious technical problem of the photovoltaic (PV) power plant (PVP) operation is a strong dependence of power generation on the plant's geographical location and climate conditions. Therefore, a reliable prediction of the electrical power generation is crucial for the PVP design to enable the appropriate choice of the plant equipment that ensures the energy and economic efficiency of the PVP.

The data on the average daily values of solar radiation intensity at the installation site is primarily used to predict the plant power output:

$$W = L\eta_1 \eta_2 S N , \tag{1}$$

where W is the expected power output per day [kWh/d)]; L is the average daily value of solar radiation intensity per unit area for every month of the year [kWh/m^2/d]; η_1 is the PV module efficiency; η_2 is the overall efficiency of power conversion; S is the PV module area [m^2]; N is the number of PV modules in a solar panel.

The solar radiation intensity value in equation (1) can be determined either by calculating according to the known methods [1–3] or by using the meteorological databases [4–6]. The obvious advantage of this approach is its maximum simplicity, but it does not take into account the actual operating conditions of the PV module operation.

In addition to the solar radiation intensity, the PV converter energy characteristics are affected by the PV module surface temperature, which in turn depends on the ambient temperature, electric load and wind speed. In the regions of Russia that have continental climates, the ambient temperature can vary over a wide range according to seasons of the year or within 24 hours that can lead to significant errors in the power output prediction.

The research aims to develop an improved method to predict the plant power output under actual operating conditions.

2. The object of the Study and the Method Requirements

The object of the study is a stand-alone PVP, its typical scheme is shown in Figure 1.

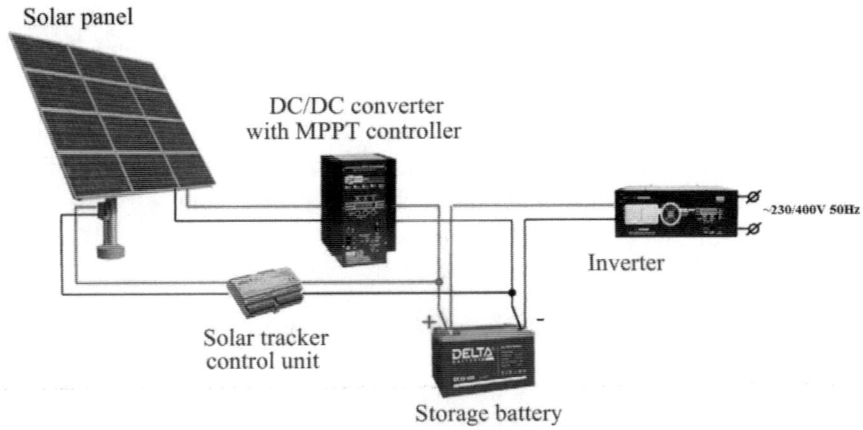

Figure 1. Scheme of the PV power plant.

The main elements of the PVP are the solar panel that consists of PV modules, DC converter controlled by the maximum power point tracking (MPPT) controller, the energy storage device, and the voltage output inverter. The PVP is often provided with the sun tracking system to improve its energy efficiency. In this case, the PVP also includes the solar tracker control unit and the tracking system operation unit based on the engines with gearboxes.

The main power equipment consistency requires careful coordination of the energy production and consumption modes. Thus, it is crucial to predict the hourly solar radiation arriving on the surface that is arbitrarily oriented to the sun and to determine the PV module energy characteristics with due regard to the external climatic factors. Non-linearity of the main PVP element characteristics and stochastic nature of the external meteorological factors significantly complicate the stated problem solution. This results in the necessity to use the mathematical modeling techniques.

The consistency of the individual components within a single system is essential to develop the PVP model, and therefore a single universal simulation tool is required. The mathematical package MatLab was used to run a simulation. The models of all the PV system components were implemented in Simulink as separate subsystems that enabled their use for simulation and study of the PVP modes of arbitrary configuration.

3. Mathematical Models of the PVP Individual Components

3.1 Model of Solar Radiation Arrival

The primary solar energy value available for conversion depends on the total solar radiation intensity at the installation site, geographical location of the plant, the solar panel spatial orientation, and the external meteorological factors: cloudiness level and surface reflection coefficient.

Most of the meteorological factors have a stochastic nature, and their time fluctuations vary from fractions of a second up to at least tens of thousands of years that eliminates the possibility of their reliable determination by the theoretical calculation methods. Therefore, the probabilistic (statistical) methods of calculating the radiation characteristics are currently used to design PVPs. The climatic manuals or electronic databases that provide the results of long-term weather observations are used as the initial data.

The total intensity of solar radiation hitting the surface, which is inclined and arbitrarily oriented relative to the tilt β and azimuth γ angles, is calculated according to the method proposed by Liu-Jordan [2], and the impact of cloudiness is determined by the Berlyand formula [7].

The initial simulation data are the PVP location latitude, azimuth and tilt angles of the solar panel orientation, number of the simulated day of the year, cloudiness level, surface albedo, and solar tracker type.

The model enables determination of the time values of solar insolation for any day of the year in any geographical location of the PVP. The theoretical method is used to calculate the insolation value for ideal conditions (clear sky). The insolation under cloudy conditions is calculated using statistical data of the meteorological databases and with due regard to the scattered radiation reflected from the earth's surface [4–6].

The model is developed as a subsystem that consists of eight separate functional units: data input unit, units for calculation of the sun height and azimuth, solar time of sunrise (sunset), solar radiation hitting the inclined area, the incidence angle of direct radiation, and units for the cloudiness level estimation and the output data preparation.

The long-term statistical data of solar radiation intensity [MJ/m^2] on a horizontal surface in some regions of Russia located between latitudes 40°N and 60°N were used to verify the model. The performed numerical experiments showed that the solar radiation arrival model provides the adequate results of calculation of the solar energy resources: the root mean square error for individual months does not exceed 14% and is no more than 8% for the whole year.

3.2. Solar Panel Model

The majority of the solar panel models are based on mathematical equations that describe the processes in the equivalent circuit of a solar cell [8, 9]. In this case, the number of unknown variables in the resulting system of equations is greater than the number of equations. Therefore, the numerical methods with a small iteration step size are essential to solve the resulting system. This type of the solar panel models is primarily used for the analysis of dynamic processes in the PVP, but the models are inefficient to calculate the solar panel energy characteristics over long periods of time.

The method proposed by Jones and Underwood [10] was used to develop a simplified model of the solar panel. According to the method, the maximum output power of the solar panel is determined by:

$$P_{\text{out}} = C_{FF} \, N_{\text{m}} \, \eta_{\text{conv}} \frac{G \ln(10^6 G)}{T}, \tag{2}$$

where N_{m} is the number of PV modules in the solar panel; C_{FF} is a constant coefficient of the solar panel; η_{conv} is efficiency of the converter with the maximum power controller; G is the current level of global insolation, W/m^2; T is the PV module current temperature.

The present study uses a more accurate equation to determine the coefficient C_{FF} in contrast to the original method [10]:

$$C_{FF} = \frac{FF\,T_{\text{ref}}}{G_{\text{ref}}} \frac{\left[I_{SC} + k_I\left(T - T_{\text{ref}}\right)\right]\left[V_{OC} + k_V\left(T - T_{\text{ref}}\right)\right]}{\ln(10^6\,G_{\text{ref}})}, \tag{3}$$

where FF is the fill factor of the PV module; T_{ref}, G_{ref} are respectively the temperature and global insolation values of the PV module under reference conditions.

The fill factor of the PV module is determined according to its technical specifications:

$$FF = \frac{I_{MPP} V_{MPP}}{I_{SC} V_{OC}}, \tag{4}$$

where I_{MPP}, V_{MPP} are certified values of the PV module current and voltage at the maximum power point under reference conditions; I_{SC}, V_{OC} are certified values of the PV module short-circuit current and open-circuit voltage under standard conditions.

The simplified model of the solar panel was implemented as a subsystem and verified on a number of PV modules. The model showed a good agreement of the simulation results with the manufacturer's data. The initial data for calculation of the PV module characteristics are the current values of solar radiation, the semiconductor surface operating temperature, and the PV module parameters listed in its technical specifications.

3.3. Model for Calculating the PV Module Surface Temperature under Operating Conditions

The solar panels located in the high northern latitudes of Russia are affected by the external meteorological factors, and therefore these factors must be taken into account. It is fairly simple to determine the average monthly ambient temperature and wind speed, which are independent from the designed installation parameters. The statistical data of meteorological observations that can be obtained from meteorological site archives are the initial data for determination of the air temperature and wind speed [11]. In contrast to the wind speed, changes in the diurnal temperature range in the northern latitudes can be determined by:

$$T(t) = \overline{T} + \frac{\Delta T}{2} \cdot \cos\left[\frac{2\pi}{t_v} \cdot (t_{loc} - t_{max})\right], \tag{5}$$

where $\overline{T}$ is the average daily air temperature, [°C]; ΔT is the diurnal air temperature variation, [°C]; t_v is the air temperature variation period, [h]; t_{max} is local time of the temperature maximum, [h]; t_{loc} is local solar time, [h].

The proposed method to develop the diurnal temperature range model can be easily implemented for any day of the year and any region using the statistical data of meteorological observations from the nearby weather station. The model is implemented as a subsystem and enables us to obtain time dependence of the temperature variation for the given day of the year using only 3 parameters (Figure 2,a).

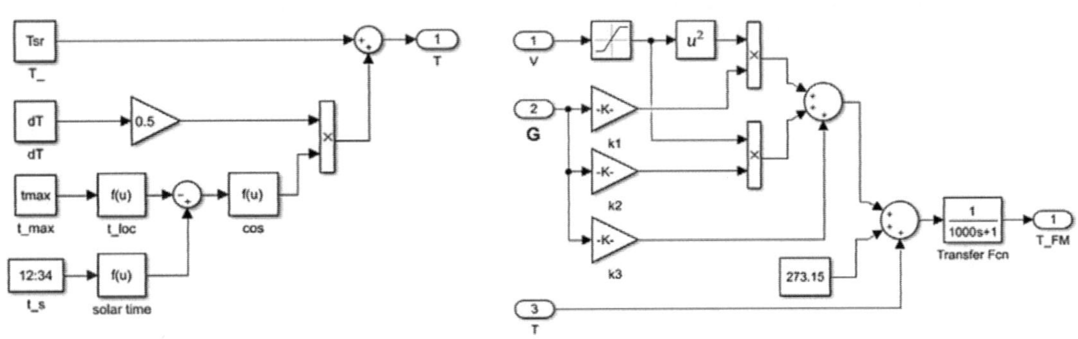

Figure 2. Model of the diurnal temperature range and model for the PV module temperature calculation implemented in MatLab/Simulink.

However, in addition to the ambient temperature impact, the PV module temperature is directly affected by the radiation energy arriving at the module surface along with the sunlight and the internal

IOP Conf. Series: Materials Science and Engineering **189** (2017) 012008 doi:10.1088/1757-899X/189/1/012008

heat sources that occur due to electrical losses in the solar cells and contact resistances. The PV module operating temperature is the result of a convective/radiant heat exchange between the module surface and the environment. Thus, the module current temperature cannot be determined by rigorous theoretical methods due to the impact of numerous factors: coefficients of the heat transfer and the absorption of module material radiation, the wind strength and direction, ratio of the face and rear surfaces of the module, etc. The results of the PVP operation have shown that the excess of the PV module temperature over that of the environment at high insolation values can reach 30°C. Therefore, the PV module actual temperature under operating conditions should be definitely considered. The empirical relationships obtained under actual operating conditions are used due to the complexity of theoretical determination of the module temperature. The empirical dependence of the PV module surface temperature excess $\Delta T_{FM} = f(G,V)$ on solar radiation intensity G and wind average speed V was obtained as a result of processing of the experimental data on the silicon PV modules operation under actual operating conditions in Siberia. The resulting dependence was implemented in MatLab/Simulink as a model for calculating the PV module surface temperature according to the given values of the average wind speed V, solar radiation intensity G and ambient temperature T, as shown in Figure 2,b. Since the PV module temperature cannot change instantly due to the thermal inertia, the inertial unit implemented on the basis of the transfer function unit was included in the model.

4. Calculation of the Predicted Power Output from the PVP

Consider a practical implementation of the proposed method to calculate the predicted power output from the PVP located in Tomsk (geographical coordinates 56ºN, 85ºE). The PVP scheme that includes the solar panel consisting of 12 PV modules of polycrystalline type JAP6-60-260 with nominal power of 260 W and the storage battery based on 12 gel lead acid batteries Delta GX 12-200 was analyzed. The initial data for simulation are the area latitude, average daily wind speed at the installation site, cloudiness level, surface albedo, number of the day of the year, azimuth and vertical angles of the solar panel installation, solar tracker type, average daily air temperature and its daily amplitude, local time of the temperature maximum, and PV module specifications. The secular trends of solar radiation intensity, diurnal air temperature range, current temperature of PV modules and their energy characteristics: short-circuit current, open-circuit voltage, fill factor, and output electric power are determined in the process of simulation. The parameters to be recorded are as follows: solar insolation value E, W/m^2; output power of the solar panel P, W; PV module and air temperature T, °C. The measured interval is one day (86 400 seconds) with a sampling interval of 1 second. The oscillograms of the recorded parameters obtained through the computational experiment that simulates the PVP operation on 20th June are shown in Figure 3.

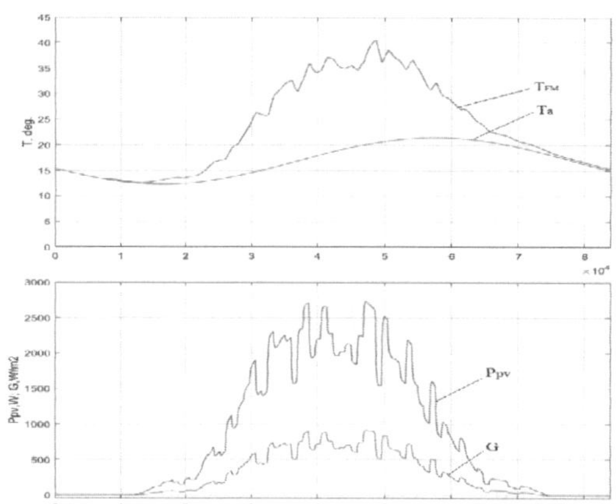

Figure 3. Oscillograms of the parameters obtained through the computational experiment that simulates the PVP operation on 20th June.

5. Conclusion

The original simulation model of the photovoltaic power plant has been developed to simulate and investigate the plant operating modes under actual operating conditions. The proposed model considers the impact of the external climatic factors on the solar panel energy characteristics that improves accuracy in the power output prediction.

The data obtained through the PVP operation simulation enable a well-reasoned choice of the required capacity for storage devices and determination of the rational algorithms to control the energy complex. In addition, obtained results will be demanded in the study of autonomous hybrid wind-solar power systems on the basis of invertor-type diesel generator sets [12]. The use of the developed photovoltaic plant model will allow to create a fully functional simulation model of autonomous power supply system with real power equipment [13, 14].

References

[1] Klein S A 1977 *Solar Energy* **19(4)** 325–329 DOI: 10.1016/0038-092X(77)90001-9
[2] Liu B Y H *and* Jordan R C 1961 *ASHRAE Journal* **3** 53–59
[3] Duffie J A *and* Beckman W A 1991 *Solar engineering of thermal process* (New York: John Wiley and Sons)
[4] *Surface meteorology and Solar Energy* Retrieved from https://eosweb.larc.nasa.gov/sse/
[5] *Irradiation data for every place on Earth* Retrieved from http://www.meteonorm.com/
[6] *RETScreen Expert* Retrieved from http://www.retscreen.net
[7] Sivkov S I 1968 *Methods for calculating the characteristics of solar radiation* (Leningrad: Gidrometeoizdat) (in Russian)
[8] Bonkoungou D *et al* 2013 *IJETAE* **3(3)** 493–502
[9] Obukhov S G *et al* 2016 *IOP Conf. Ser.: Mater. Sci. Eng.* **132(1)** 012017 DOI: 10.1088/1757-899X/132/1/012017
[10] Jones A D *and* Underwood C P 2001 *Solar Energy* **70(4)** 349–359 DOI: 10.1016/S0038-092X(00)00149-3
[11] *Weather in 243 Countries of the World* Retrieved from http://www.rp5.ru/
[12] Obuhov S *et al* 2016 *MATEC Web of Conferences* **91** 01036 DOI: 10.1051/matecconf/20179101036
[13] Lukutin B V *et al* 2016 *IOP Conf. Ser.: Mater. Sci. Eng.* **127(1)** 012005 DOI: 10.1088/1757-899X/127/1/012005
[14] Lukutin B V *et al* 2014 *Proc. of the Int. Conf.* EEEIC'14 459-463 doi:10.1109/EEEIC.2014.6835913

Using the Semiconductors Materials of InSb-ZnTe System in Sensors for Gas Control

E G Shubenkova

Associate Professor, Omsk State Technical University, Omsk, Russia

E-mail: shubenkova-e@mail.ru

Abstract. The samples of thin film semiconductor compounds InSb, ZnTe and solid solutions based on them were obtained by vapor deposition of components on a dielectric substrate in a vacuum, followed by annealing and their surface properties in CO, O_2 and NH_3 gas atmospheres were investigated. Identification of the samples was carried out by X-ray diffraction techniques. In the temperature range 253 ÷ 403 K and a pressure range of 1÷12 Pa the gas adsorption was measured by piezoelectric microbalance technique. In order to establish the basic regularities of processes flowing on samples surface in addition to the electrophisical were used Infrared and Raman spectroscopic measurements. The resulting addiction "surface property - composition" is extreme and have allowed to determine solid solution $InSb_{0,95}$-$ZnTe_{0,05}$ as the most sensitive to the presence of ammonia, selective and this sample exhibits a negligible oxidation of surface.

1. Introduction

Using of semiconductor sensors allows not only to do rapid detection and analysis of trace element in different technological environments, as well as check the harmful components of gas emissions from motor vehicles, chemical and petrochemical industries and the surrounding polluted areas.

In comparison to other methods of analysis, semiconductor sensors cause interest with their relative cheapness, compact size, ultra-high sensitivity and high reliability. Analytical sensor signal is caused by change in conductivity or other characteristics of the semiconductor surface during adsorption on it particles of different nature. Herewith, the sensor signal is proportional to the concentration of the particles.

Influence of adsorption processes on the resistance of the semiconductor material is explained by depletion of the conduction band relative to the electrons in the case of adsorption of oxidizing gases (O_2, CO_2, etc.), which leads to an increase of the energy barrier in the contact area of the particles and, as a result, the resistance of gas sensitive element increases. In other case the number of electrons at the surface of contact area increases in the case of gas adsorption with reducing properties (H_2, CO, etc.), the resistance of gas sensitive material is reduced accordingly.

Among the sensory analyzers are actively studied semiconducting oxides, but one of the most significant drawbacks is their influence on their characteristics in an open atmosphere, meteorological parameters such as temperature and humidity. Therefore, the attention of researchers to films and film structures A^3B^5, A^2B^6 compounds as devoid of such defects quite justified. Binary compounds and their solid solutions show themselves as promising materials for control of gas media [1–6].

Using the A^3B^5, A^2B^6 type compounds and solid solutions based on them as components of the sensor element provides a sensitivity to a broad range of toxic and chemically-active gases in the environment and gas emissions, but it should be noted that for each gas exists a fairly narrow temperature range, in which the sensitivity of the material to it the highest [7].

The basis of described semiconductor sensors are thin (250 nanometers) layers InSb, ZnTe and their solid solutions deposited on appropriate substrates (glass, potassium bromide, etc.). Selection of sensitive material for express diagnostics and control represents obtain samples of different composition, and studying the sensitivity of the electrical conductivity of the surface in adsorption different gases as well as the dynamics of the sensor signal formation.

2. Experimental

In the polycrystalline film samples InSb, ZnTe and their solid solutions in a concentration up to 20 mol.% ZnTe in InSb was prepared by depositing the vapor of components on a dielectric substrate in a vacuum with following annealing.

The radiographic studies were performed on BRUKER AXS Advanced D8 Powder X-ray Diffractometer instrument (CuKα-radiation, λ=1.54056 Å, T = 293 K) using the high-angle scattering technique [8].

The electron-microscopic studies were performed on JCM-5700 scanning election microscope instrumented with JED 2300 energy-dispersive analysis attachment.

Adsorption of CO, O_2, NH_3 in the temperature range of 253÷403 K and pressure range 1÷12 Pa was measured by piezoelectric microbalance method.

Electrical and IR and Raman spectroscopic measurements were used in addition to direct adsorption studies to establish the basic dependencies and mechanisms of interaction the surface with gases of different nature (CO, O_2, NH_3). IR absorption spectra were recorded on a spectrophotometer InfraLUM FT-02.

The experimental data, their reproducibility and accuracy was monitored by parallel measurements, and the use of computational algorithms of mathematical statistics.

3. Results and considerations

On the results obtained by x-ray diffraction studies films of binary components and solid solutions crystallized in a cubic structure. The presence and shift of corresponding lines in the diffraction patterns of thin films showed the formation of solid solutions InSb-ZnTe system. The dependence of the unit cell values and radiographic density is linear (figure 1), which is generally characteristic of the solid solutions.

The conducted optical researches allowed to determine the the band gap of the thin film samples (Eg), which has the following values: for InSb is 0.212 eV; $(InSb)_{0.95}$-$(ZnTe)_{0.05}$ is 0.285 eV; $(InSb)_{0.90}$-$(ZnTe)_{0.10}$ is 0.378 eV.

The specific surface conductivity with increasing amount of ZnTe in InSb up to 20 mol.% demonstrates a decrease exponentially due to the substitution of corresponding elements atoms in the crystal lattice of starting components (Figure 1). The format of the dependencies of the specific surface conductivity and the band gap of InSb-ZnTe system (Figure 1) allows us to conclude about the formation of solid solutions at these compositions.

The surface of the films A^3B^5, A^2B^6 and their solid solutions show significant heterogeneity, which is manifested in their contact with the gas media, and the microscopic study of surface films. Surface defects, defects in the crystal lattice, cluster formations ranging in size from a few to two or three dozen nanometers. However, the reproducibility of the synthesis conditions may reproducible physical and chemical properties of thin film samples. The presence and location of defects on the surface, their size, geometry of gas molecules affect the sensitivity and selectivity of thin film samples in relation to the different components of the gaseous media.

The magnitude of adsorption gases lies within $10^{-3} \div 10^{-5}$ mol/m^2. The experimental dependence of the adsorption on the temperature was allowed to specify the intervals reproducible operation of the film surface, which corresponds limits of a reversible chemical adsorption [8–9].

On the basis of the experimental relationships "adsorption value - thin film composition", samples with appreciable sensitivity to the studied gases were obtained. Namely, in the case of ammonia such adsorbent proved solid solution $InSb_{0.95}$-$ZnTe_{0.05}$. On a sample of this composition at the same external parameters value of adsorption all studied gases is two or three times lower compared with NH_3.

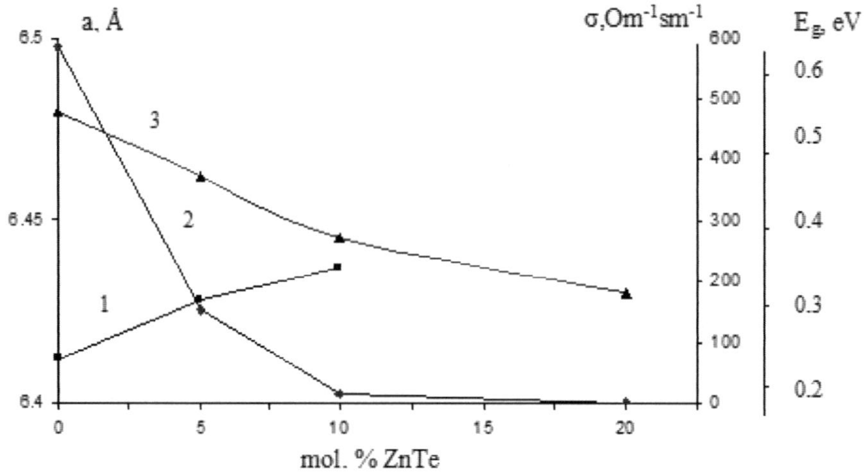

Figure 1. The dependences of the optical band gap (1), conductivity (2) and a cell period (3) of the system InSb - ZnTe components.

The analysis dependences of electrical properties of the studied sample surface from contact with the polluting gas is a necessary research step. Dynamics of changes surface conductivity relatively the vacuum in medium of CO, O_2 and NH_3 was studied by probe method on piezoelectric resonator (Figure 2, 3). The solid solution with composition $InSb_{0.95}$-$ZnTe_{0.05}$ in study showed the highest sensitivity of the surface characteristics to the presence of ammonia and carbon monoxide.

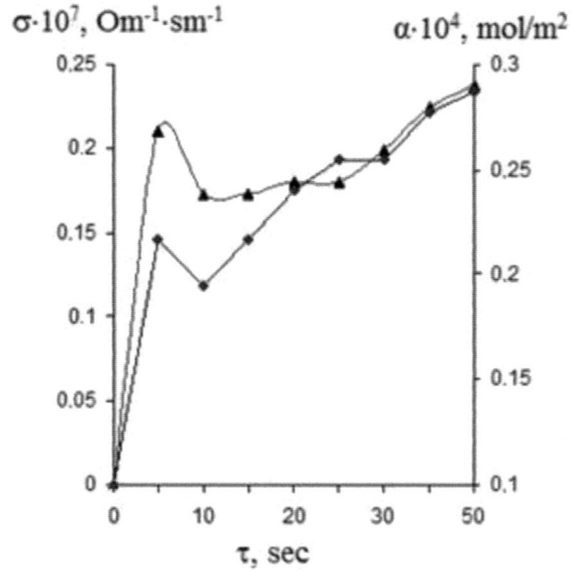

Figure 2. Dependences of the electrical conductivity of the surface (1) and the value of adsorption (2) for a contact time of $InSb_{0.95}$-$ZnTe_{0.05}$ sample in CO atmosphere (p = 5.3 Pa).

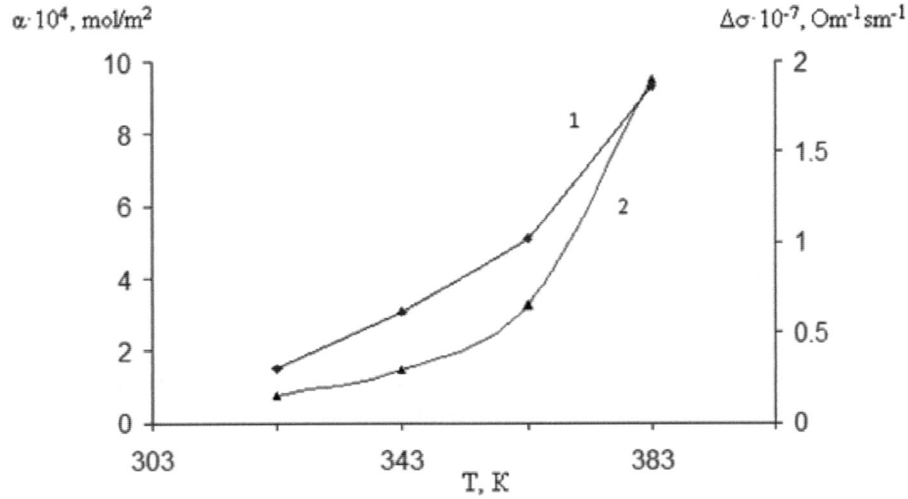

Figure 3. Dependences of the change the surface conductivity (1) and the value of adsorption (2) for $InSb_{0.95}$-$ZnTe_{0.05}$ sample in NH_3 atmosphere (p = 5.3 Pa).

That results correspond to the results of the adsorption measurements and confirmed their relationship for the semiconductor surface [8-9].

4. Summary

Thus, by method of evaporation in a vacuum the thin film samples of semiconductor A^3B^5, A^2B^6 compounds and solid solutions based on them with reproducible physical and chemical properties are obtained.

These samples are identified by X-ray diffraction techniques.

The surface adsorption, electrical and optical properties of InSb-ZnTe system components are studied.

Dependences of "surface property - composition" have an extreme view and reveals the optimum composition of the solid solution containing $InSb_{0.95}$-$ZnTe_{0.05}$, surface electrical conductivity of which is most strongly changes in ammonia, has more selectivity and a low surface oxidizability.

Due to possibility of regulating the surface properties combined with high surface activity and nano-scale irregularities present on the surface, the thin films of InSb-ZnTe system can be recommended as active elements for sensors.

References

[1] Cattrall R W 1997 *Chemical Sensors* (New York: Oxford University Press Inc.)
[2] Yamazoe N *et al* 1996 *Mater. Sci. Eng., B* **41 (1)** 178–181 DOI: 10.1016/S0921-5107(96)01648-0
[3] Dai Z R *et al* 2003 *Adv. Funct. Mater.* **13(1)** 9–24 DOI: 10.1002/adfm.200390013
[4] Simon I *et al* 2001 *Sensors and Actuators, B* **73(1)** 1–26 DOI: 10.1016/S0925-4005(00)00639-0
[5] Moshnikov V A 2011 *Inorg. Mater.* **47(1)** 18–22 DOI: 10.1134/S0020168510121039
[6] Moshnikov V A *et al* 2010 *J. Non-Cryst. Solids* **356(37-40)** 2020–2025 DOI: 10.1016/j.jnoncrysol.2010.06.030
[7] Isaienko O *et al* 2005 *Sensors and Actuators, B* **108(1-2)** 134–142 DOI: 10.1016/j.snb.2004.12.081
[8] Kirovskaya I A *and* Shubenkova E G 2009 *Russ. J. Phys. Chem.* **83(13)** 2322–2330 DOI: 10.1134/S003602440913024X
[9] Kirovskaya I A *et al* 2010 *Russ. J. Phys. Chem.* **84(4)** 661–667 DOI: 10.1134/S0036024410040242

Injection Process Control of the Well at the Hydrodynamic Research of Coalbed

I G Odnokopylov[1], O V Galtseva[1], I Yu Krasnov[1], A O Smirnov[1], M S Karpov[3], O A Surzhikova[4], V V Kuznetsov[5], and J Li[6]

[1]Associate Professor, National Research Tomsk Polytechnic University, Tomsk, Russia
[2]Senior Lecturer, National Research Tomsk Polytechnic University, Tomsk, Russia
[3]Student, National Research Tomsk Polytechnic University, Tomsk, Russia
[4]Associate Professor, Czech Technical University in Prague, Czech Republic
[5]Associate Professor, Omsk State Technic University, Omsk, Russia
[6]Professor, Tsinghua University, Beijing, China

E-mail: odivan@yandex.ru

Abstract. This scientific work is devoted to the study results of water injection process into the well at the hydrodynamic research by using the high pressure unregulated pump. The injection process should be accompanied by the retention of some hydraulic parameters at constant level during some time. Various variants for use of mechatronic nodes for automatization of water injection process are considered. Scheme for reducing the load on the pump and equipment in hydraulic system and also for improving the quality control system with high accuracy is shown. Simulation results of injection process into the well at the pressure and consumption fixation and recommendations for the use of the proposed schemes depending on the technological process are given.

1. Introduction

Development of oil and gas fields is based on the modern scientific technologies. They are built on detailed studying of properties of productive layers, gases and fluids and the processes in layers during operation. Research process of fields includes the stages from their opening to full depletion of recoverable gas and oil reserves. This information is used for control, designing and management by field development systems for ensuring the paces and reserve degree of depletion at a minimum cost.

Hydrodynamic (industrial) researches of layers and wells actively are used for obtaining of the necessary information [1]. At the same time the characteristics of the studied layers are specified; the possibility of effective impact on a bottomhole zone of layer is revealed. Hydrodynamic simulation results allow to choose the intensification technology of gas inflow and to carry out the choice of suitable mode of equipment operating of the wells.

2. Hydrodynamic Research of the Wells

Hydrodynamic experiments are implemented by fluid injection into the layer during certain period of time with registration of wellhead and bottomhole pressures. It is known that the relative permeabilities are taken into account by using the technologies of the hydrodynamic studies at the multiphase filtration flow. Therefore, the filtration flow in the crack system is single-phase for simplification of the data interpretation of such tests [2–7].

The scheme of underground equipment of well is used at conducting of the hydrodynamic studies; its scheme is shown in Figure 1.

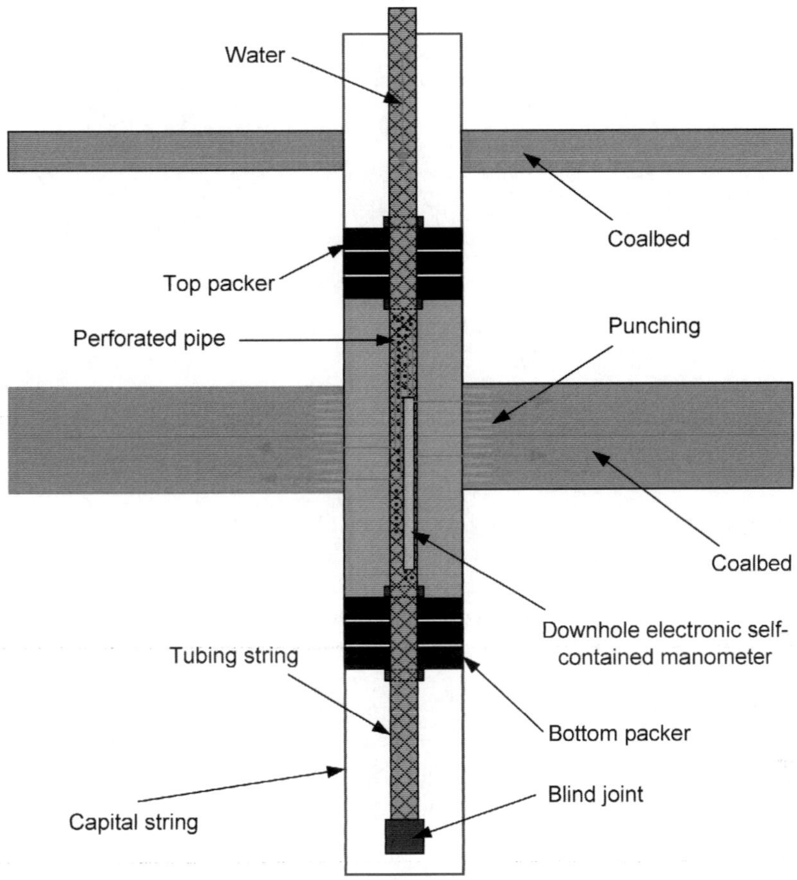

Figure 1. Scheme of underground equipment of well at hydrodynamic researches.

It is necessary to conduct data interpretation of hydrodynamic tests for evaluating productive and filtration characteristics of the layers and wells (pressure of layer, productivity or filter coefficients, permeability, skin factor, etc.) and features of the borehole and remote zones of layer. The quality of the results interpretation depends on the compliance accuracy of the hydrodynamic research technology [8–11]. It is required to ensure steady monitoring of wellhead and bottomhole pressure and for supporting a constant flow fluid during pumping in the presence of disturbances in the near-well area [12–16].

3. Realization of Fluid Injecting into Well

The maintaining of defined consumption of fluid and pressure in the well is done by the operator by using the valve during a long time at the fluid injecting into a well at the hydrodynamic studies.

It was proposed the use of regulating high pressure pump for fluid injecting automation with high precision. The pump outlet is connected via the sensors directly to the well. The main objective of this solution is the creation of deep regulating pump which can hold a small consumption at high pressure with high precision. This decision is not considered in this paper.

The fluid consumption is proposed to be controlled through the regulating throttle.

Figure 2 shows the proposed variant of the process automation scheme for research wells by using the through-type regulating valve.

The work of the present scheme is described below. The fluid from tank (1) is supplied to the unregulated high-pressure pump (3) of low pressure pipeline (2); further, on the high-pressure pipeline fluid is supplied via the regulating valve (4), which regulates the amount of fluid coming into the well. The regulating valve is driven by asynchronous motor (6) through gear (5).

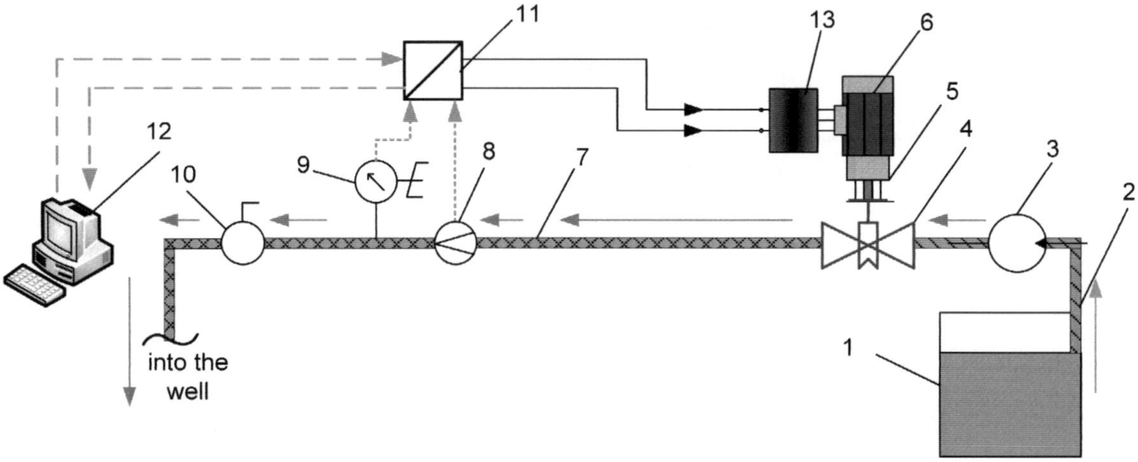

Figure 2. Functional automation scheme by using the through-type regulating valve.

After that the fluid moves through the flowmeter (8) into the well through the high-pressure pipeline (7). Data from the flowmeter (8) and pressure sensor (9) is processed by programmable logic controller (11); the formed task is transmitted to frequency converter (13). Personal computer (12) is used for visually displaying the information in real time and recording the data in defined format. The pressure pipeline is closing by using a ball valve (10) after reaching the needed pressure in the well. The injection process is complete.

Figure 3 shows a functional scheme of the hydrodynamic studies by using the through-type regulating valve in the dumping letdown. The work of the present scheme is described below. The working fluid is in the tank (1); it goes into the unregulated high-pressure pump (3) on the low-pressure pipeline (2). In the pipeline (4) under a high pressure fluid is supplied to the T-bend (5), which divides the fluid flow into the working and waste flow.

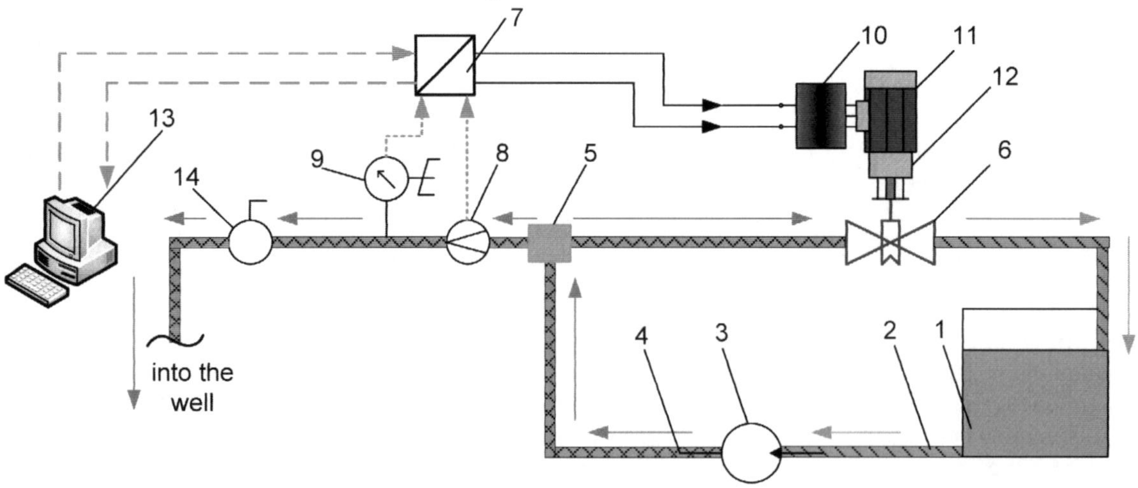

Figure 3. Functional scheme of the hydrodynamic studies by using the through-type regulating valve in the dumping letdown.

After the T-bend a working flow passes through the flowmeter (8) and goes in the well. The selected pump productivity is more than the required consumption of working fluid injection into the

IOP Conf. Series: Materials Science and Engineering **189** (2017) 012010 doi:10.1088/1757-899X/189/1/012010

well, so some part of the working fluid goes on the return pipeline through the regulating valve (6) into the container (1). Consumption of working fluid injection into the well on defined level can be maintained by changing the amount of fluid which is discharged back into the tank.

Control of the working fluid injecting consumption into the well provides by the programmed logic controller (7). This controller receives data from the flowmeter and pressure sensor (9); after processing data it forms the task for frequency converter (10) which feeds the induction motor (11). Next through a reducing gear (12) the engine drives the stem of the regulating valve (6), which varies the quantity of passing fluid. The personal computer (13) is required for visual display of information in real time and for recording the data into the nonvolatile memory. Electronic manometer (9) is needed to control the pressure in the well. It is necessary to close the injection pipeline by using the ball valve (14) once the well pressure will reached a defined level.

According to analysis of the both variants the scheme (Figure 4) has many disadvantages such as the high cost and low efficiency of the system. This is because most of the fluid which supplied by the pump (3) is discharged back into the tank (1). However, the scheme allows making the injection process with high accuracy; electrical and hydraulic equipment for realization is accessible.

4. Simulation of the of Injection Fluid Process into the Well

Simulation model of the injection fluid into the well has been developed in the Matlab software. For simulation in statics and dynamics the SimPowerSystems library (for modeling of adjustable speed drives) [10] and the SimHydraulics library (for modeling of hydraulic systems) are used. The model scheme includes a hydraulic tank (for storage of working fluid), a pump (for injecting the working fluid in the system); regulating valve (throttle); fluid conductors (for connection of hydraulic elements to each other). Appropriate sensors in the fluid conductor are used to measure pressure and consumption. Water is used as working fluid; losses in the fluid conductors were not taken into account. Figures 4, 5 show the simulation results of the hydrodynamic studies with the use of through-type regulating valve in supply pipeline and the dumping letdown.

Figure 4 shows the transient curves of P pressure during the starting up the pump and the regulating the valve in supply pipeline (red curve in Figures 2, 3) and in the dumping letdown (blue curve in Figures 2, 3).

Figure 5 shows the transient curves of Q consumption during the starting up the pump and the regulating the valve in supply pipeline (red curve in Figures 2, 3) and in the dumping letdown (blue curve in Figures 2, 3).

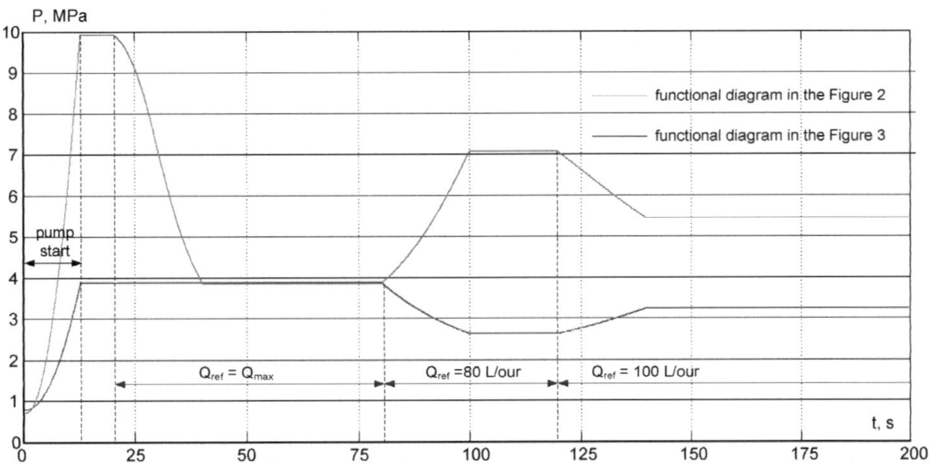

Figure 4. Transient curves of P pressure during the starting up the pump and the regulating the valve.

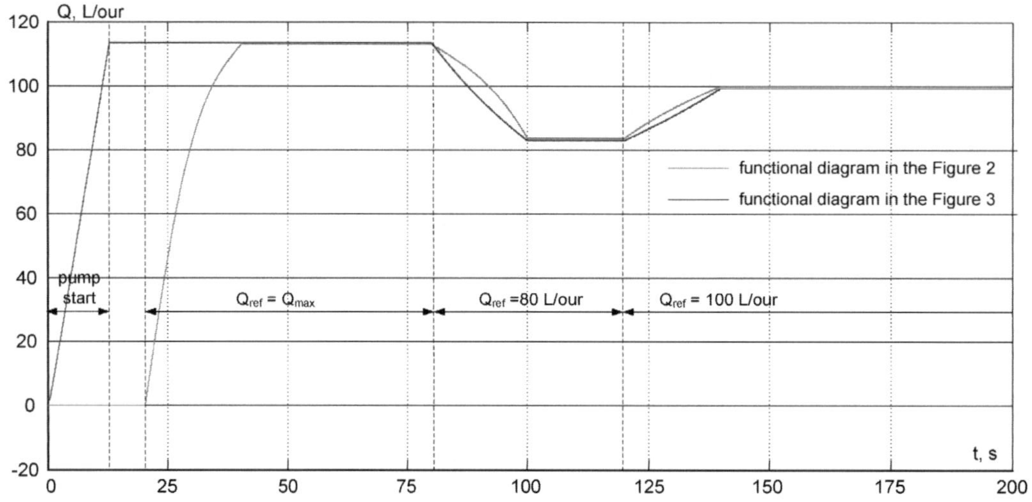

Figure 5. Transient curves of Q consumption during the starting up the pump and the regulating the valve.

Transient curves of consumption for both variants are almost identical except the stage of the pump starting up. If through-type regulating valve is installed in the delivery pipe, the beginning of fluid injection into the well is determined by the position of the valve; in this case the injection into the borehole starts at 20 seconds. If through-type regulating valve is installed in the dumping letdown, the beginning of fluid injection is determined by the moment of the pump starting up.

Analyzing Figure 4, it is possible to say that the fluid consumption regulation is characterized by elevated pressure in the dumping letdown at the implementation of the first variant of placement of the through-type regulating valve. It creates a strain on the pump and elements of the hydraulic system that will reduce the equipment life.

5. Conclusion

Method of injection fluid into the coalbed for permeability determination is well established in low-permeability fractured-porous reservoirs of the CBM fields. It allows determining the filtration parameters of the coalbeds more reliably for a short time at a low cost.

Based on the data interpretation experience of the hydrodynamic studies, the most successful tests are at the constant fluid consumption during the pumping period throughout the all test. Interpreted data analysis of the hydrodynamic studies shows that the fluid consumption stabilization on the defined level throughout the all test is a mandatory condition for obtaining of the acceptable data.

Conducted research of various automated schemes of the hydrodynamic tests show that the using the through-type regulating valve is a most simple and effective way of the accurate maintenance of the working fluid consumption.

Practice shows that the fluid consumption control is performed more precisely at the low pressures; the variant with through-type regulating valve in the dumping letdown is more preferred.

References

[1] Saulsberry J L *et al* 1996 *Gas Research Institute Report GRI-94/0397* (Chicago: Gas Research Institute)
[2] Hopkins C W *et al* 1998 *Proc. of the SPE Gas Technology Symposium'98* 9–24
[3] Negrov D A *et al* 2017 *IOP Conf. Ser.: Mater. Sci. Eng.* **168(1)** 012061 DOI: 10.1088/1757-899X/168/1/012018
[4] Yakimov E *et al* 2017 *J. of Physics: Conf. Ser.* **783** 012061 DOI: 10.1088/1742-6596/783/1/012061
[5] Surzhikova O A 2015 *IOP Conf. Ser.: Mater. Sci. Eng.* **81(1)** 012098 DOI: 10.1088/1757-899X/81/1/012098
[6] Skutin E D *et al* 2016 *Procedia Engineering* **152** 493–496 DOI: 10.1016/j.proeng.2016.07.633
[7] Langraf S *et al* 2017 *MATEC Web of Conferences* **102** 01026 DOI: 10.1051/matecconf/201710201026

[8] Vlasov V A *et al* 2015 *IOP Conf. Ser.: Mater. Sci. Eng.* **71** 012029 DOI: 10.1088/1757-899X/71/1/012029
[9] Taco G *et al* 2010 *SPE Asia Pacific Oil and Gas Conference and Exhibition* **2** 1240–1245
[10] Galtseva O V *et al* 2016 *IOP Conf. Ser.: Mater. Sci. Eng.* **110** 012094 DOI: 10.1088/1757-899X/110/1/012094
[11] Krasnov I *et al* 2015 *IOP Conf. Ser.: Mater. Sci. Eng.* **93** 012061 DOI: 10.1088/1757-899X/93/1/012061
[12] Odnokopylov I G *et al* 2015 *Proc. of Int. Conf. SIBCON'15* DOI: 10.1109/SIBCON.2015.7147249
[13] Frantsina E V *et al* 2016 *AIP Conference Proceedings* **1772** 060017 DOI: 10.1063/1.4964597
[14] Mustafina R M *et al* 2016 *J. of Physics: Conf. Ser.* **671** DOI: 10.1088/1742-6596/671/1/012046
[15] Iuriev I Y *et al* 2015 *IOP Conf. Ser.: Mater. Sci. Eng.* **71** 012022 DOI: 10.1088/1757-899X/71/1/012022

Simulation of Industrial Wastewater Treatment from the Suspended Impurities into the Flooded Waste Mining Workings

L Bondareva[1], Yu Zakharov[2], and A Goudov[2]

[1]Senior Lecture, Kemerovo State University, Institute of Fundamental Sciences Kemerovo, Russia
[2]Professor, Kemerovo State University, Institute of Fundamental Sciences, Kemerovo, Russia

E-mail: l.v.kemerova@mail.ru

Abstract. The paper is dedicated to the mathematical model of slurry wastewater treatment and disposal in a flooded mine working. The goal of the research is to develop and analyze the mathematical model of suspended impurities flow and distribution. Impurity sedimentation model is under consideration. Due to the sediment compaction problem solution domain can be modified. The model allows making a forecast whether volley emission is possible. Numerical simulation results for "Kolchuginskaya" coal mine presented. Impurity concentration diagrams in outflow corresponding to the real full-scale data obtained. Safely operation time mine workings like a wastewater treatment facility are estimated. The carried out calculations demonstrate that the method of industrial wastewater treatment in flooded waste mine workings can be put into practice but it is very important to observe all the processes going on to avoid volley emission of accumulated impurities.

1. Introduction

Coal industrial wastewater treatment is still a key and very serious problem for Kuzbass. Slurry water composition is varied. It may contain suspended particles (coal dust, rock dust, clay particles), salts of heavy metals, phenols, ammonia, nitrates, nitrites, free sulfuric acid, sulfur and other hazardous substances. Their emissions into water objects can cause changes of their irreversible environmental impact. That's why wastewater should be treated or disposed.

Currently, treatment facilities of the coal enterprises, along with the traditional wastewater treatment [1–4] technology implemented another approach - requires treatment process with the help of flooded waste mining workings [5]. Natural purification of wastewater is supposed to take place in mine workings due to the water precipitation and mixture with influent underground water. This method is applied to purify slurry water of "Komsomolec" coal preparation plant in the waste mine workings of "Kolchuginskaya" coal mine.

This alternative approach is of great concern. Although, the method is quite simple and requires low costs it is essential to be researched and to be controlled underground treatment processes. In particular, it is important to avoid a volley emission when the pollution in the pumped fluid can increase greatly and significantly. Flooded mine workings can be defined to be a black box, that's why mathematical simulation and numerical experiments is the most convenient to investigation treatment processes.

At present most of papers devoted to model of the spread of suspended impurities, related to the study sediment processes [6], the formation of sediment in reservoirs or at the seaside [7–8], spreading and transport oil slicks [9–10].

There are no models allows to define how features of the sludge treatment from suspended impurities, deposition of sediments and fluid flow can affect on wastewater treatment quality in flooded mine workings and to estimate safely operation time mine workings like a wastewater treatment facility. In this paper, the suggested mathematical model considered all this factors. Mathematical and numerical simulation of wastewater treatment of dissolved impurities is described in [5].

The goal of the research is to develop and analyze the mathematical model of suspended impurities flow and distribution, and to estimate safely operation time mine workings like a wastewater treatment facility.

2. Experimental

We shall examine the solution domain G with boundary ∂G that represents flooded mine working form [5]. We consider fluid to be uniform, viscous and incompressible. Nondimentional Navier–Stokes equations system describes this fluid flow by variables "flow function-vortex"

$$\frac{\partial \omega}{\partial t} + u\frac{\partial \omega}{\partial x} + \upsilon\frac{\partial \omega}{\partial y} = \frac{1}{\mathrm{Re}}\Delta\omega;$$

$$\frac{\partial^2 \psi}{\partial x^2} + \frac{\partial^2 \psi}{\partial y^2} = -\omega.$$

with certain initial conditions

$$u|_{t=0} = 0, \quad \upsilon|_{t=0} = 0, \quad \omega|_{t=0} = 0, \quad \psi|_{t=0} = 0;$$

and boundary conditions

$$U = U_1(t,x,y); \quad \omega|_{\partial G} = \left(\frac{\partial \upsilon}{\partial x} - \frac{\partial u}{\partial y}\right)_{\partial G}; \quad \psi|_{\partial G} = \psi_1(t,x,y),$$

where $\vec{U} = (u(t,x,y), \upsilon(t,x,y))$ – velocity vector, defined by its components u, υ; ω – vortex, ψ – flow function; $U_1(t,x,y)$, $\psi_1(t,x,y)$ – known functions, estimated on the solution domain boundary ∂G; Re – Reynold's number.

To simulate distribution of impurities transfer equation is

$$\frac{\partial C}{\partial t} + u\frac{\partial C}{\partial x} + (\upsilon - \upsilon_S)\frac{\partial C}{\partial y} = D\left(\frac{\partial^2 C}{\partial x^2} + \frac{\partial^2 C}{\partial y^2}\right)$$

with certain initial $C(x,y,0) = C_0(x,y)$ and boundary conditions. The concentration in the enter fluid flow $C_1(x,y)$ and on the entire upper boundary $C_2(x,y)$ are known and constant, the condition of this type $\frac{\partial C}{\partial x} = 0$ are put on the outlet boundary, and on the lower boundary we put boundary condition $\frac{\partial C}{\partial y} = \alpha C$.

IOP Conf. Series: Materials Science and Engineering **189** (2017) 012011 doi:10.1088/1757-899X/189/1/012011

Here C is concentration of impurity sedimentation, υ_s is velocity of impurity sedimentation, D is diffusion coefficient, α is the coefficient determines the intensity of sediment accumulation at the bottom.

Subsidence impurity and slumped impurity can eventually affect bottom shape. We simulate this process in the following way: if during the time interval T^* settled impurity concentration exceeds threshold value C^* near the boundary of solution domain, then this impurity is considered not to be affected by the flow and the solution domain boundary is shifted according to concentration C^* and time T^*.

3. Results and considerations

The formulated differential problems are solved by the grid method. The vorticity transfer equation and the impurity transfer equation are solved by the implicit scheme of stabilizing corrections. Poisson difference equation for flow function is solved by the minimal residual method of not full approximation with matrix parameter by applying component wise and overall optimization of iteration parameters [5].

Figure 1 shows the dynamics of impurity distribution and sedimentation when time values t = 1000 months for (1) impurity "heavy fraction", (2) impurity "light fraction", (3) total fractions.

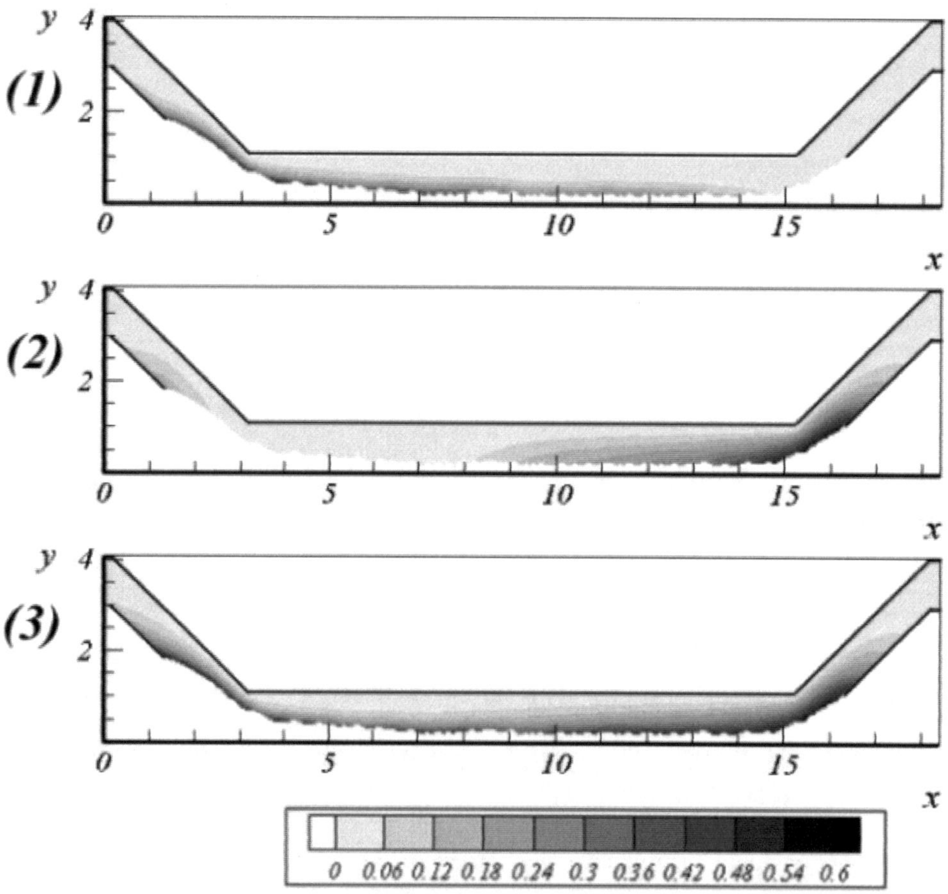

Figure 1. Two fractions particle impurity flow and distribution.

Impurity "heavy fraction" is determined with the following parameters' values $C_1 = 0.025$; $C_2 = 0$; $D = 1$; $\upsilon_S = 2.5$; $\alpha = -5$; $C* = 0.6$; $T* = 1$. Impurity "light fraction" parameters are $C_1 = 0.025$; $C_2 = 0.00003$; $D = 4$; $\upsilon_S = 0.1$; $\alpha = -1$; $C* = 0.9$; $T* = 10$. The results for two fractions particle impurity from the "heavy" and "light" fractions are presented in (Figure 1). All nondimentionall parameters values correspond to the full-scale data obtained for the "Kolchuginskaya" coal mine [5].

Particles of "heavy" fractions are large enough to be quickly deposited and compacted sediment, it is almost not diffusion affected. Having these particles determine the intensity of the compacted sediments mine waste filling. The "light" fraction availability affects the turbidity of wastewater and has a significant impact on the outflow impurity concentration, which is shown in diagram (Figure 2).

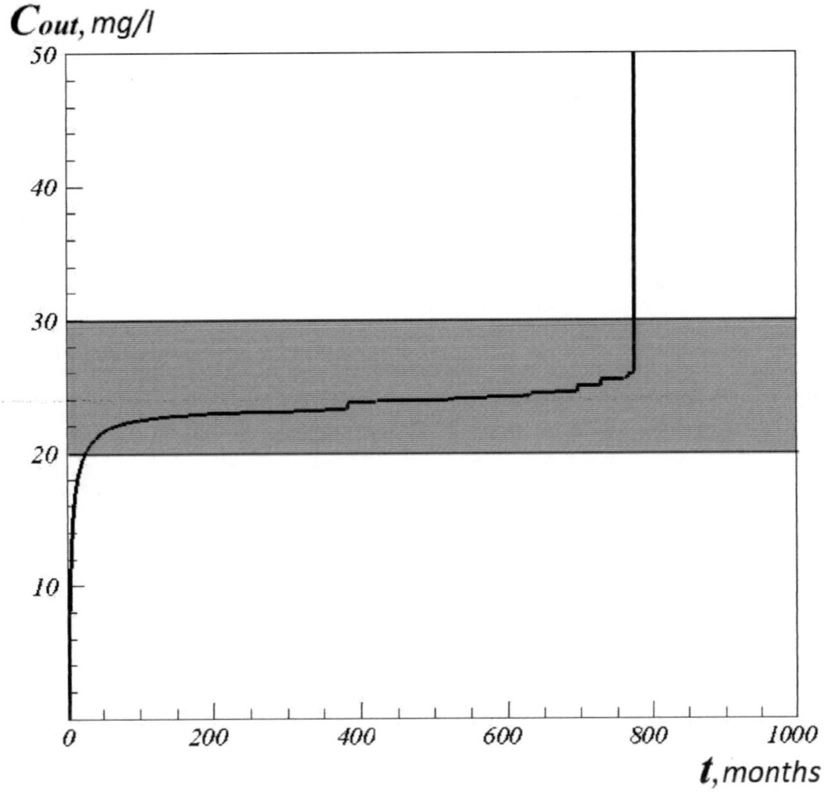

Figure 2. Diagram of impurity quantity change leaving the solution domain. A range of values corresponding to the full-scale data obtained for the "Kolchuginskaya" coal mine grayed.

Because of, these data are consistent with "Kolchuginskaya" coal mine full-scale data we can conclude that the waste water disposal process will be "safe" for approximately 60 years while maintaining constant treatment conditions. However, after this time-out, waste mine workings can be filled significantly by compacted sediment. This can lead to impurity volley emission, when the impurity concentration in outflow substantially increases.

4. Summary

The mathematical model of suspended impurity flow, distribution and settling enables to change solution domain due to the deposition of sediments in flooded mine workings suggested in this paper. Numerical simulation results for "Kolchuginskaya" coal mine presented. Impurity concentration

IOP Conf. Series: Materials Science and Engineering **189** (2017) 012011 doi:10.1088/1757-899X/189/1/012011

diagrams in outflow corresponding to the real full-scale data obtained. In more details, model and obtained using it numerical simulation results is described in [11].

Acknowledgements
The research was carried out with support of the project part state task 1.630.2014/K and of the base part state task 2014/6.

References

[1] Hudson-Edwards K *and* Dold B 2015 *Minerals* **5(1)** 82–85 DOI: 10.3390/min5010082
[2] Environmental Protection Agency 2014 *Reference guide to treatment technologies for mining-influenced water* (U.S.: EPA)
[3] Galtseva O V *et al* 2016 *IOP Conf. Ser.: Mater. Sci. Eng.* **132** 012003 DOI: 10.1088/1757-899X/132/1/012003
[4] Punkkinen H 2016 *Guidelines for mine water management* (Finland: Julkaisija – Utgivare – Publisher)
[5] Zakharov U N *et al* 2013 *Simulation of impurities distribution in flooded mine workings: monography* (Kemerovo: Publishing House of the KemSU) (in Russian)
[6] Saletti M *et al* 2016 *Earth Surf. Dynam.* **4** 549–566 DOI: 10.5194/esurf-4-549-2016
[7] Carretier S *et al* 2016 *Earth Surf. Dynam.* **4** 237–251 DOI: 10.5194/esurf-4-237-2016
[8] Van Maren D S *et al* 2015 *Continental Shelf Research* **95** 1–14 DOI: 10.1016/j.csr.2014.12.010
[9] Lubartseva S 2016 *Nat. Hazards Earth Syst. Sci.* **16** 2009–2020 DOI: 10.5194/nhess-16-2009-2016
[10] Aditya K M *and* Suresh K G 2015 *Aquatic Procedia* **4** 435–442 DOI: 10.1016/j.aqpro.2015.02.058
[11] Zakharov Yu *and* Bondareva L 2015 *Procedia Engineering* **117** 389–396 DOI: 10.1016/j.proeng.2015.08.183

IOP Conf. Series: Materials Science and Engineering **189** (2017) 012012 doi:10.1088/1757-899X/189/1/012012

Determination of the Optimal Position of Pendulums of an Active Self-balancing Device

G R Ziyakaev [1], O A Kazakova [1], V V Yankov [2], and O P Ivkina [3]

[1] Associate Professor, National Research Tomsk Polytechnic University, Tomsk, Russia
[2] Postgraduate student, National Research Tomsk Polytechnic University, Tomsk, Russia
[3] Master's student, National Research Tomsk Polytechnic University, Tomsk, Russia

E-mail: vladimir_2894@mail.ru

Abstract. The demand of the modern manufacturing industry for machines with high motion speed leads to increased load and vibration activity of the main elements of rotor systems. Vibration reduces operating life of bearings, has adversary effects on human organism, and can cause accidents. One way to compensate for a rotating rotor's imbalance is the use of active self-balancing devices. The aim of this work is to determine the position of their pendulums, in which the imbalance is minimized. As a result of the study, a formula for determining the angle of the pendulums was obtained.

1. Introduction

At present, introducing new technologies and increasing the efficiency of manufacturing processes require the design of high-performance machines with high motion speeds, high performance reliability and precision. One of the main components of modern machines are high-speed rotating elements (rotor systems) with a high load capacity and vibration activity. The presence of vibration leads to heavy wear of bearings, and, in some cases, decreases the accuracy of the performance of a device's or machine's functions. Vibration has a direct impact on humans, decreasing one's performance.

Modern methods and ways of balancing of rotating rotors allow to balance them to the high class of accuracy. In some cases, the initial balancing during operation of the device or machine is compromised and does not meet the requirements. For example, in industrial fans this happens due to uneven carbon buildup on the blades. In different centrifuges, change of imbalance occurs very quickly and with every start. Therefore, for such types of machines, automatic rotor balancing becomes necessary.

Presently, machine designers use active self-balancing devices (SBDs) for spinning rotors. They allow to decrease instrument and equipment vibration, decrease the wear of bearings and other rotating parts. As a result, the operating life of instruments and machinery increases. Self-balancing devices are divided into 2 types: active and passive. In passive devices, balancing weights move freely under the effect of internal forces, but in the active ones, this happens by means of some outside force [1]. Both of them have their advantages and disadvantages. The main advantage of passive SBDs is their simplicity. They do not require external power supply and complex control schemes, so they are usually quite compact and easy to manufacture. But the main disadvantage of passive SBDs is that they decrease the vibration only at speeds above critical, i.e. above resonance. In the area below resonance, such devices only increase the imbalance of the system and worsen the resonance transition [2-4]. Active SBDs do not have this disadvantage – they work at all rotor speeds, but their structure is complex and they need complex control systems [5-7].

Since the balancing process in the active SBDs does not depend on the rotor speed, and the speed and trajectory of the balancing weights are determined by the imbalance and depend on the kinematic model, the type of the balancing weights displacement drive and the sensitivity of the measuring system, it is obvious that active SBDs can achieve high accuracy of the installation of the balancing weights in the positions that ensure the maximum possible balance of the rotor in accordance with the hardware configuration. Active SBDs are based on the principle of closed-loop automatic control systems and consist of a large number of individual blocks that perform different functions. Eventually this leads to increased complexity of such devices, their bulkiness and high cost.

Active SBDs differ in the method of searching for positions of the balancing weights on SBDs with random search and SBDs with guided movement of the balancing weights. Devices with random search are constructed so that random search provides independence of their work from the rotor speed. The disadvantage of such system is that the imbalance can increase indefinitely [5], and balancing time is not constant due to the random nature of the search. However, the balancing process may be made more reliable if we introduce some regularity in the random search method.

Balancing devices with guided movement of the balancing weights do not require the measure of the imbalance phases and can operate with a measuring system that shows only the presence of vibration in supports and the change in their amplitude. This measuring system is simpler and more reliable because it has less components and functions.

Figure 1 shows the basic scheme of an active SBD. They work on the following principle: a vibration sensor measures the vibration in the machine's casing or support and feeds the data to a controller, where it is processed. The controller gives control signals to drive mechanisms, which set balancing weights relative to the rotor in such a way that they compensate for its imbalance and dampen the vibration.

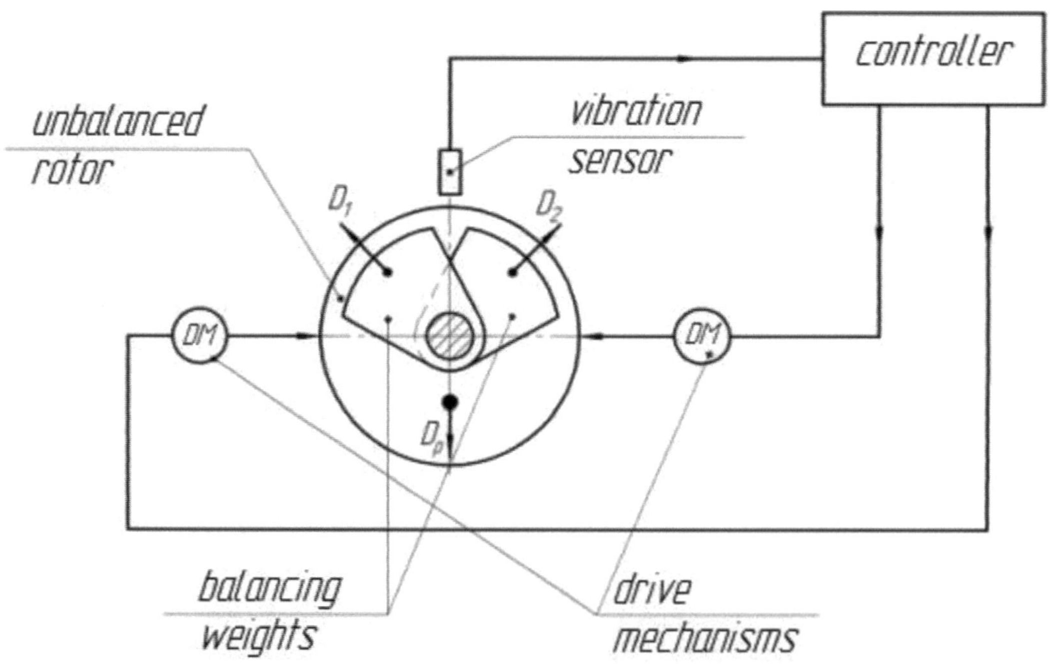

Figure 1. Basic scheme of an active SBD

2. Determination of the pendulums' position

Because the main vector of the system imbalance and the balancing weights (pendulums) are in different planes, complete equilibration of the system cannot be achieved. Consequently, dynamic reactions will appear in rotor supports. The aim of this work is to find the position of the pendulums, in which the influence of the residual imbalance is minimal.

Let us consider a mechanical system in Figure 2. A rotor with an unbalanced disc has the imbalance D and rotates on the supports A and B, the distance between which is l. Balancing weights, namely the pendulums with the imbalance Dm, deviate from the vector of the disc imbalance at the same angle φ. The distance between the pendulums is c, the distance between the support and the central SBD plane is a, the distance between the central plane of the SBD and the disc is b.

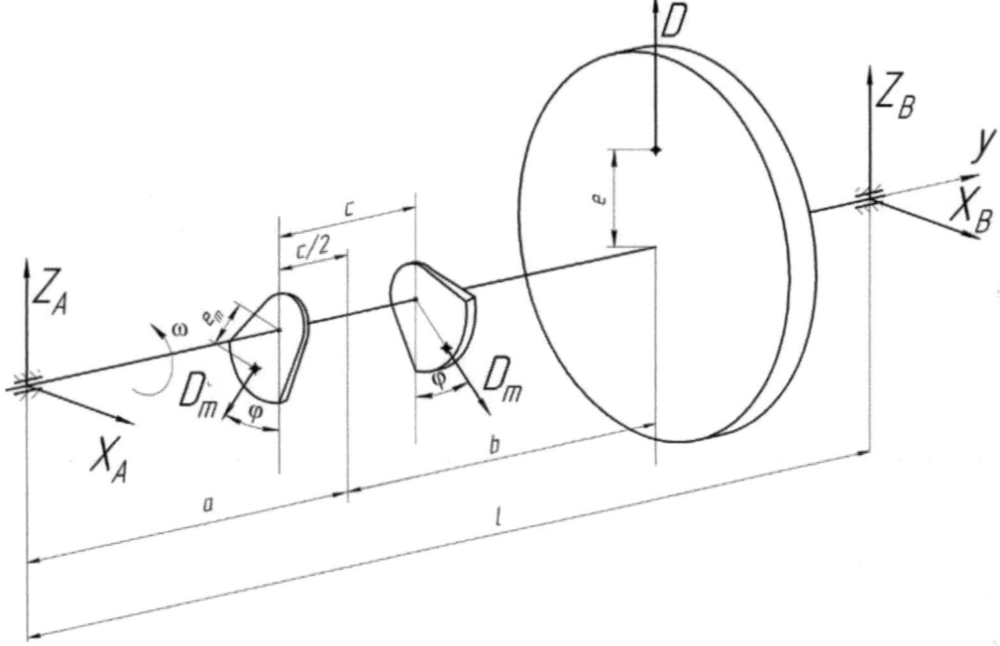

Figure 2. Basic scheme of an active SBD.

In terms of the equilibrium conditions, the components of the reactions in the bearings are equal to:

$$Z_A = \frac{2\,D_m \cos\varphi\,(l-a) - D\,(l-a-b)}{l};$$

$$Z_B = \frac{2\,D_m \cos\varphi\, a - D\,(a+b)}{l};$$

$$X_A = X_B = \frac{D_m \sin\varphi\, c}{l};$$

Accordingly, the overall reactions are:

$$R_A = \sqrt{\left(\frac{2\,D_m \cos\varphi\,(l-a) - D\,(l-a-b)}{l}\right)^2 + \left(\frac{D_m \sin\varphi\, c}{l}\right)^2};$$

$$R_B = \sqrt{\left(\frac{2D_m \cos\varphi\, a - D(a+b)}{l}\right)^2 + \left(\frac{D_m \sin\varphi\, c}{l}\right)^2};$$

The objective of the research is to find the angle φ, at which the dynamic reactions in the bearings are minimal:

$$F = \left(R_A(\varphi) + R_B(\varphi)\right) \to \min.$$

We can consider the function containing the squares of the reactions to simplify calculations:

$$F = \left(R_A^{\,2}(\varphi) + R_B^{\,2}(\varphi)\right) \to \min.$$

The squares of the reactions are:

$$R_A^{\,2} = \left(\frac{2D_m \cos\varphi(l-a) - D(l-a-b)}{l}\right)^2 + \left(\frac{D_m \sin\varphi\, c}{l}\right)^2;$$

$$R_B^{\,2} = \left(\frac{2D_m \cos\varphi\, a - D(a+b)}{l}\right)^2 + \left(\frac{D_m \sin\varphi\, c}{l}\right)^2;$$

In order to find the extremum of the function, let us take the derivative and set it equal to zero:

$$\frac{d(R_A^{\,2} + R_B^{\,2})}{d\varphi} = 0;$$

After the transformations, we get:

$$\frac{\left(-16a^2 - 8l^2 + 16al + 4c^2\right)D_m^2 \cos\varphi \sin\varphi + \left(-8la + 4l^2 + 8a^2 - 4lb + 8ab\right)D_m D \sin\varphi}{l^2} = 0;$$

or

$$\left(-8a^2 - 4l^2 + 8al + 2c^2\right)D_m^2 \sin(2\varphi) + \left(-8la + 4l^2 + 8a^2 - 4lb + 8ab\right)D_m D \sin\varphi = 0;$$

By solving the equation for φ, we finally obtain:

$$\varphi = \mathrm{arctg}\left(\frac{\sqrt{-\left(-8la + 4l^2 + 8a^2 - 4lb + 8ab\right)^2 D_m^2 D^2 + 4\left(-8a^2 - 4l^2 + 8al + 2c^2\right)^2 D_m^4}}{\left(-8la + 4l^2 + 8a^2 - 4lb + 8ab\right)D_m D}\right)$$

3. Conclusions

Active SBDs of different types allow to achieve high-precision balancing and performance at all frequencies of rotor rotation. However, these devices are complex. The higher the accuracy of the device, the more complex and more perfect the elements should be. These systems are huge, massive, and, most importantly, expensive. In is reasonable to apply this type of balancing devices in equipment where the technological process requires high balancing accuracy and where it is economically feasible to use high-precision measuring systems and drives.

In recent years, there has been a further development of active balancing devices, measuring systems, automated corrective mass movement control systems, as well as improvements in hardware. This study resulted in obtaining a formula for determining the angle of SBD pendulums, in which total dynamic reactions in the bearings are minimal. This allows to continue improving the properties of active SBDs in the simplest, most reliable and effective way, i.e. by improving their design, kinematic layout, which in turn are determined by the drive type.

References

[1] Ziyakaev G R *et al* 2001 *Proc. of RUSKO-AM* 223–226
[2] Moon J – D *et al* 2006 *International Journal of Machine Tools and Manufacture* **46(9)** 978–987 DOI: 10.1016/j.ijmachtools.2005.07.048
[3] Bu J *et al* 2011 *Advanced Materials Research* **287-290** 2906–2910 DOI: 10.4028/www.scientific.net/AMR.287-290.2906
[4] Hredzak B *et al* 2003 *Proc. of PEDS* 8046602 DOI: 10.1109/PEDS.2003.1283179
[5] Hredzak B *and* Guo G 2006 *IEEE Transactions on Magnetics* **42(5)** 8888668 DOI: 10.1109/TMAG.2005.863619
[6] Svetashkov A A *et al* 2016 *Key Engineering Materials* **685** DOI: 10.4028/www.scientific.net/KEM.685.206
[7] Kupriyanov N A *et al* 2016 *IOP Conf. Ser.: Mater. Sci. and Eng.* **124** 0121100 DOI: 10.1088/1757-899X/124/1/012100
[8] Orszulik R *and* Shan J 2012 *Smart Materials and Structures* **21(5)** 055002 DOI: 10.1088/0964-1726/21/5/055002
[9] Hredzak B *and* Guo G 2004 *Proc. of Instn Mech. Engrs* 218(C) 401–410 DOI: 10.1177/095440620421800405
[10] Zhou S *and* Shi J 2002 *J. of Sound and Vibration* **249(1)** 196–205 DOI: 10.1006/jsvi.2001.3660
[11] Bagutdinov R A *and* Narimanov R K 2015 Proc. of European Conference on Informational Technology and Computer Science 17–22
[12] Ivkina O P *et al* 2016 *J. Phys.: Conf. Ser.* **744** 012157 DOI:10.1088/1742-6596/744/1/012157
[13] Blanco-Ortega A *et al* 2012 *Advances on Analysis and Control of Vibrations - Theory and Applications* **10** (InTechDOI: 10.5772/48621
[14] Hong Z K *et al* 2014 Applied Mechanics and Materials **620** 304–309 DOI: 10.4028/www.scientific.net/AMM.620.304

IOP Conf. Series: Materials Science and Engineering **189** (2017) 012013 doi:10.1088/1757-899X/189/1/012013

Bottom Ash Waste Used in Different Construction Materials

G G Volokitin[1], N K Skripnikova[1], O G Volokitin[2], A V Lutsenko[2], V V Shekhovtsov[3], V A Litvinova[2], and M A Semenovykh[4]

[1]Professor, Tomsk State University of Architecture and Building, Tomsk, Russia
[2]Associate Professor, Tomsk State University of Architecture and Building, Tomsk, Russia
[3]Graduate Student, Tomsk State University of Architecture and Building, Tomsk, Russia
[4]Student, Tomsk State University of Architecture and Building, Tomsk, Russia

E-mail: nks2003@mail.ru

Abstract. The paper presents the investigations of the composition and properties of the bottom ash waste generated by Seversk thermal power plant of the Tomsk region. The compositions suggested for construction materials (fired and hydrothermal) are based on the bottom ash waste. Mechanical-and-physical and physicochemical properties of the produced specimens are investigated in this paper. Phase compositions are determined for fired materials based on the bottom ash waste, and a possibility of producing ash-based mineral wool is defined herein. The produced ash-containing fiber possesses a higher chemical resistance and performance characteristics due to its higher acidity index.

1. Introduction

The industrial production of wall materials such as ceramic brick and different types of concrete, faces the increasing lack in high-quality raw materials the finished products of which suit the market demands [1]. The respective literature represents a variety of investigations in the field of ash applications in constriction, although its practical use is occasional. Therefore, a rational utilization of ash waste is still relevant to the construction materials production, at a regional level, in particular [2].

The paper aims to give an assessment of the possible addition of ash waste to the mixture compositions intended for the production of ceramic brick and different types of concrete. With this view, the experiments are conducted for the selection of the appropriate raw materials composition. Ash waste is prepared in to stages: air-drying at 50°C and ball milling.

2. Experimental

Clay raw material from Verkhovoe deposit of the Tomsk region was used in this experiment as a binder component. Ash waste was prepared in to stages, namely: air-drying at 50°C and ball milling.

The dry compaction technique is used to prepare ceramic bricks from Verkhovoe deposit clay. Specifications for the procedure include 10÷12% molding-moisture content; 25 MPa molding pressure; 24-hour dying; 950°C burning temperature. The ash content in the mixture ranges from 10 to 100 wt.% [3].

Ash waste is characterized by the high content of Al_2O_3 which results in the formation of mullite-like compounds when interacted with silica during burning. Iron oxides (~ 5%) facilitate the formation of a large amount of the primary melt and complex compounds [4].

The grain size distribution is one of the main indicators of raw materials. The more the content of fine particles the higher the plasticity of the raw material. Therefore, the latter possesses a higher coherency that has a positive effect on the strength properties of finished ceramic products. Taking all this into account, the ash waste is subjected to different grades of milling. The grain size distribution for the ash waste is given in Table 1.

Table 1. Grain size distribution for the ash waste.

Raw material, [%]	Grain size, [μm]					
	<3	3–5	5–7	7–10	10–14	>14
Clay	84.4	7.4	3.3	2.4	1.5	1
Ash-1	84.1	6.9	3.0	2.2	1.5	2.3
Ash-2	92.3	4.1	2.0	1.4	0.2	—

As can be seen from Table 1, the grain size distribution of ash is close to that of clay. A high-performance milling allows producing ash powder with 92.3% content of fine particles (<3 μm).

The obtained specimens are subjected to physicochemical testing that provides measurements of density, strength, water absorption, and frost resistance.

The increase of the ash content in the mixture from 25 to 100% results in the decrease in density which ranges from 1900 to 1300 kg/m^3. The compressive strength as one of the basic properties of ceramic brick varies between 33.5–11.0 MPa.

The X-ray diffraction (XRD) analysis of different brick specimens and physicochemical processes occurred in ceramics burning is carried out by DRON-3 diffractometer. The XRD analysis allows observing the formation of crystal phases, while the scanning electron microscope (SEM) investigations identify phase formations by their specific properties. These techniques provide an understanding of the structure formation mechanisms in a ceramic specimen in each case and give an idea of the regulation of mechanical-and-physical properties of ceramic products [5].

The XRD analysis is carried out for ceramic specimens made of ash from Seversk thermal power plant (TPP). The XRD results are compared with the XRD patterns of 100% clay and ash raw materials. Such major crystal phases as quartz (0.335; 0.425 nm), anorthite (0.320; 0.251 nm) and mullite (0.269; 0.228 nm) are found in clay specimens burned both at 975 and 1000 C. Helenite phases in the ash-containing mixture vanish due to the low content of off-grade ash (40 wt.%). It should be noted that the burning-out of the organic component is high as well as the burning temperature (1000°C), that facilitates the phase formation and improves the quality of the primary melt during the thermal treatment of the specimen [6].

Figure 1 contains SEM images of different magnifications of ceramic specimens. Magnification of 1000x shows that the specimens made of the efficient ash compositions and burned at 1000°C, have a large amount of vitreous phase that indicates to the dominant processes of the liquid phase sintering involving the clay binder. Intergranular voids and a large number of voids generated by burning-out of oil-carbon sludge are visible at magnifications of 5000x and 10000x. Needle-shaped crystals are shown at 40000x magnification. These crystals are embedded in the vitreous phase that proves the presence of mullite-like compounds in the ceramic composition that contains clay raw materials and aluminum silicate waste.

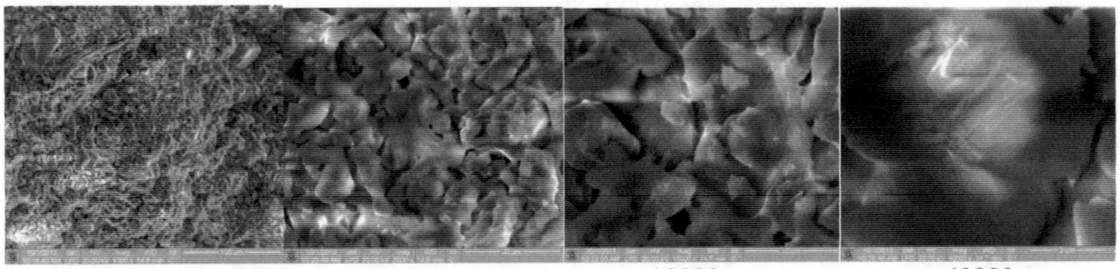
1000x 5000x 10000x 40000x

Figure 1. SEM images of ceramic specimens of the optimum composition (60% clay, 40% ash from Seversk TPP).

Research results demonstrate the following phase transformations in the mixture after the burning ceramic specimens containing ash waste:

- The primary melt is formed from the clay binder and some of low-melting ash grains;
- Such strengthening phases as mullite- and anorthite-like compounds are formed.

Cavity healing and shrinkage of unmelted aluminum silicate particles with the following formation of a monolithic product are the main structural modifications that strengthened the ceramic specimens [7].

The obtained specimens are subjected to mechanical-and-physical tests in compliance with the SNiP 530–2007 and SNiP 8462–85 taking into account the scaling factor for a single solid brick.

The use of ash waste in the production of ceramic brick requires the introduction of a batch box in the process chain. In accordance with the joint operation agreement between Tomsk State University of Architecture and Building and LLC Sibdom, this technology is being successfully implemented at this construction company.

The use of ash waste obtained during the production of heavy and fine concretes is one more promising direction of the ceramic brick production.

The introduction of the optimum amount of ash in concrete mixture improves its pouring, lowers water permeability and concrete shrinkage and enhances its frost resistance. The addition of ash waste has no a negative effect on the elasticity modulus of concrete and at the same time, increases its sulphate resistance.

According to the regulatory documents available in Russia, ash waste can be used to prepare concrete mixtures.

A choice of concrete compositions containing ash waste is conditioned by the component ratio including ash, which allows the concrete and concrete mixture to achieve the appropriate properties at the minimum cement content. Ash in a concrete mixture acts in the capacity of not only the active cementitious material, but also the micro filling material which improves the size distribution of silica sand and affects the concrete structure formation. The finest fractions of ash can act as puzzolanes. In terms of the multifunctional behavior of ash, its introduction in the mixture instead of a silica sand portion promotes the optimization of compositions [8].

Due to the low water consumption of concrete mixtures the substitution of 20% of cement by ash waste does not practically involve their shrinkage deformations at air hardening.

The ash waste introduced in the concrete mixture has different grain size distribution depending on the concrete grade. The prepared concrete mixture is poured in the form and cured for a period of one day till its setting. Afterwards, the obtained specimens are subjected to steam curing.

The optimum conditions of heat-moisture treatment are to be considered for the preparation of steam-cured specimens. These conditions should be selected depending on ash and cement properties. In using combined ash-containing cements high temperatures of steam curing are generally preferred.

The mechanical-and-physical properties of prepared concrete specimens are determined in accordance with SNiP 10180–90 and SNiP 12730.3–78.

Table 2 summarizes the test results for heavy and fine concretes.

Table 2. Test results for heavy and fine concretes.

Composition	Mechanical-and-physical properties					
	Average density, [kg/m^3]	Compressive strength after steam curing, [MPa]	Compressive strength after 28 days in natural conditions, [MPa]	Concrete grade (closest)	Water absorption, [%]	
					W_m	W_o
N 1 (heavy concrete)	2394	34.0	38.9	B26,5(M350)	1.7	4.1
N 2 (heavy concrete)	2210	12.0	17.6	B12,5(M150)	3.5	4.7
N 3 (heavy concrete)	2330	39.0	44.9	B30(M400)	1.0	2.7
N 4 (fine concrete)	2000	13.5	18.2	B12,5(M150)	3.8	7.6
N 5 (fine concrete)	1750	13.5	19.0	B12,5(M150)	3.3	5.9
N 6 (fine concrete)	2100	26.5	32.7	B25(M350)	0.4	0.9
N 7 (fine concrete)	1800	14.5	20.7	B15(M200)	1.8	3.3

The test results show that the grade B30 heavy concrete specimens have the maximum strength properties; water absorption by its volume and weight content is minimum and equals to 1% and 2.7%, respectively.

It has been found that ash waste in the amount not exceeding 8% of the original weight is advisable to use in the production of heavy concrete, whereas for fine concrete this value should not be over 35%.

The maximum strength properties pertinent to the grade B25 concrete relate to fine concrete with the milled ash waste addition.

IOP Conf. Series: Materials Science and Engineering **189** (2017) 012013 doi:10.1088/1757-899X/189/1/012013

A possibility is verified for adding ash waste in the mineral fiber production using high-temperature technologies.

The analysis performed for the both Russian and foreign mineral fiber productions shows that basaltic rocks is the most common raw material for the production of silicate melts. This is due to the traditional technologies applied to obtain the silicate melt the melting temperature of which does not exceed 1500°C [4]. Plasma technologies in the production of mineral fibers considerably extend the range of raw materials and give a possibility of using refractory silicate-containing materials with 1900 °C and higher melting temperature. Alongside this, ash wastes containing a large amount of SiO_2 can be utilized through their processing into mineral fiber in a plasma-chemical reactor. SiO_2-based raw material will allow producing mineral fibers having the improved physicochemical and mechanical-and-physical properties.

The raw materials for silicate melt are selected depending on its melting temperature and the effect from mineral and chemical compositions on physicochemical properties of the silicate melt.

Alongside with basaltic rocks, shale coal and bottom ash wastes produced by thermal power plants can be utilized as raw materials for silicate melt production.

The mineral fiber production in melting units (cupola furnaces, bath furnaces) is based on the use of raw material mixture compositions with the acidity index ranging between 1.2÷1.5. In nature, it is hard to found the material the chemical composition of which would satisfy the requirements for melting process both in cupola and bath furnaces in order to obtain silicate melt with the proper working viscosity and surface tension necessary for the production of mineral wool with the higher performance characteristics. Therefore, in the both cupola and bath furnaces, the appropriate raw mixture composition should contain two or more components. The chemical composition and grain size distribution of present-day furnace slags do not meet the melting requirements for cupola furnaces. So the selection of raw materials, their property classification and mixture composition is one of the basic problems of the mineral wool production in some of the economic regions.

In this connection, the use of plasma treatment technologies is rather relevant for melting ash wastes. Ash waste utilization is very important for the environmental protection, however, the traditional mineral fiber technology cannot be used due to a high melting temperature of ash waste (over 1600°C) [9].

The mixture compositions are selected experimentally to produce mineral fibers characterized by a higher chemical resistance both at a standard and a higher acidity index.

Table 3 contains the chemical and mixture compositions of raw materials and mineral fiber production using plasma treatment technology.

Table 3. Chemical and mixture compositions for mineral fiber production.

Mixture composition, [wt.%]	SiO_2	Al_2O_3	Fe_2O_3	CaO	MgO	TiO_2	R_2O	Acidity index
Ash - 47 Limestone - 53	39.66	14.52	3.37	41.05	1.13	0.30	-	1.28
Ash - 80 Limestone - 20	56.10	20.47	4.66	16.69	1.60	0.43	-	4.20
Ash - 80 Alumina - 20	49.64	37.52	4.13	4.14	1.40	1.81	1.82	15.70
Ash - 90 Broken glass - 10	64.95	21.43	4.65	5.23	2.02	0.43	2.14	11.90
Ash - 100	61.95	22.53	5.15	5.09	1.75	0.48	-	12.35

For the mineral fiber production, the original components are preliminary mixed and supplied by portions to a plasma rotating reactor. This technology allows processing different ash wastes possessing a high melting temperature. The amount of the melt obtained from fine material depends on energy, gas-dynamic, and thermal characteristics of plasma reactor. It is shown that the saturation range of melt formation provided by the supplied energy ranges between 32÷53 kW. This indicates that plasma jet used in the experiments has an optimum value of the power supply a further increase of which does not lead to the increase in the efficiency of plasma reactor.

Ash-based mineral fibers are high in silicon and aluminum oxides if compared to basalt fiber. According to data found in the literature, this fact indicates to a higher chemical and thermal resistance of these fibers. As for coarse fibers obtained in the tested melts, they form in a rather wide temperature range [10].

Depending on the operation conditions, the mineral fiber should possess such important properties as diameter, flexural strength and chemical resistance to aggressive substances. As is known, with the increase in silica content in the melting product the chemical resistance to aggressive substances also increases. Calcium, magnesium, aluminum, and iron oxides have a positive effect on the fiber tolerance, whereas alkali metal oxides significantly reduce it. A study of the fiber properties is based on the methodologies applied to glass and other fiber types.

Ash-based mineral fibers produced by plasma treatment with the subsequent melt blowing out by the airflow, meet the mineral wool standard requirements. Moreover, a higher acidity index these fibers have exerts a positive effect both on their chemical and thermal resistance. A hydrolytic resistance of these fibers is also higher. However, the mineral fiber contains a larger amount of non-fibrous inclusions and has a greater diameter of an elementary fiber.

3. Conclusions

The experiments showed that ash waste in the amount of 70%, assisted in producing a ceramic brick of M150 grade strength that conditions its use in wall construction in different types of buildings. 50% ash compositions will provide the brick production of M200 grade strength which can be used in bearing wall structures.

It was found that ash waste addition increased the concrete strength both during its natural hardening and after steam curing. The frost resistance of ash-containing concrete was decreased similar to that containing other hydraulic additions.

It was shown that the fine ash particles added to the concrete mixture is one of the most efficient ways to reduce the consumption of Portland cement in construction.

Thus, the experiments showed that ash waste generated by thermal power plants can be used in the production of mineral wool. Ash waste is a prospective raw material for the production of mineral fiber with the chemical resistance and performance characteristics improved due to its higher acidity index, provided that plasma treatment is used to obtain a chemically homogeneous melt.

References

[1] Iordanidis A *et al* 2008 International Journal of Coal Geology **73(1)** 63–73 DOI: 10.1016/j.coal.2007.02.007
[2] Vatin N I *et al* 2011 *Magazine of Civil Engineering* **4(22)** 16–21 (in Russian)
[3] Bakunov V S *and* Lukin E S 2008 *Glass and Ceramics* **65(5–6)** 193–198 (in Russian)
[4] Ooi C Y *et al* 2007 Ceramics International **33 (7)** 1171–1177
[5] Kornilov A V *et al* 2009 *Glass and Ceramics* **1** 23–25 (in Russian)
[6] Biernacki J J *et al* 2008 Fuel **87(6)** 782-792. DOI: 10.1016/j.fuel.2007.08.024
[7] Iuriev I Y *et al* 2015 *IOP Conf. Ser.: Mater. Sci. Eng.* **71** 012022 DOI: 10.1088/1757-899X/71/1/012022
[8] Surzhikov *et al* 2016 *IOP Conf. Ser.: Mater. Sci. Eng.* **110** 012002 DOI: 10.1088/1757-899X/110/1/012002
[9] Vlasov V A *et al* 2015 *J. Phys.: Conf. Series* **652** 012031 DOI: 10.1088/1742-6596/652/1/012031
[10] Khatri C 2010 *Fuel Process Technology* **91(10)**1288–1295 DOI: 10.1016/j.fuproc.2010.04.011

Process of Formation of Vacuum Polymer Films in the HF Discharge Reactor

P A Luchnikov[1], and O A Surzhikova[1]

[1]Senior Researcher, Moscow Technological University (MIREA), Moscow, Russia

E-mail: xamdex@gmail.com

Abstract. Forming processes of thin-film fluoropolymer coating in the HF discharge reactor on the surface at various technological modes of deposition are considered. The molecular and structural characteristics of the films are studied. Areas of optimum technological modes of producing of nano-sized and thin fluoropolymer films with better electro physical properties are determined.

1. Introduction

Secondary polymerization principle of products of spraying of industrial polymeric substances is used at creation of producing technology of fine and nano – sized polymer coatings [1, 2]. Decomposition of block polymer into the gas molecular atmosphere can occur under temperature influence, electron beam, laser radiation or ions of plasma of HF discharge [3-5]. In this case deposition of the polymer coating on the substrate can be carried out at conditions similar to polymerization of the polymer from the monomer.

Fluoropolymer film coatings possess unique multifunctional properties due to their high electrical and physico – chemical protective characteristics, which allows to use their in electronics devices in extreme conditions and harsh atmosphere [6, 7]. Technological process of obtaining of fluoropolymer coatings with properties, close to the industrial polytetrafluoroethylene (PTFE), requires study. Secondary structure synthesis of PTFE on the surface of the substrate requires natural selection of molecules on the surface of the fluoropolymer layer during its polymerization and control of technological modes of deposition because of significant heterogeneity of molecular composition of resulting gas atmosphere.

2. Specimens and Experimental Method

Fluoropolymer films are inflicted on substrate in vacuum chamber in cathode HF discharge reactor with a frequency of 13.56 [MHz] from the vapors of sprayed block PTFE. The film thickness was 1...10 [μm]; it was measured by layer weight on the samples by quartz resonator and by ellipsometrical method. The substrate temperature was changed in the range of 20...150°C during the deposition of films. Materials of the plates (of 14 type) of silicon, steel, copper and glass – ceramic were served as substrates.

Cathode HF discharge reactor and cathode HF discharge reactor in a heterogeneous constant magnetic field are used to form the fluoropolymer film coatings. The latter type of reactor allows forming the fluoropolymer films with desired characteristics. Figure 1 shows design scheme of HF discharge reactor of magnetron type, where 1 is a cooled round cathode target 1; 2 is a refrigerated table for specimens 4, on which is formed film; 3 is a body-electrode of the reactor; 5–7 is a magnetic system. Inhomogeneous magnetic field *B* is created by the permanent magnets 5 and 7, which is directed perpendicular to the alternating electric field between the electrodes of the capacitive cell of cathode target 1 and the second body-electrode of reactor 3 using rigidly fixed on the body of the magnetic system 5–7, located on the periphery of the HF reactor.

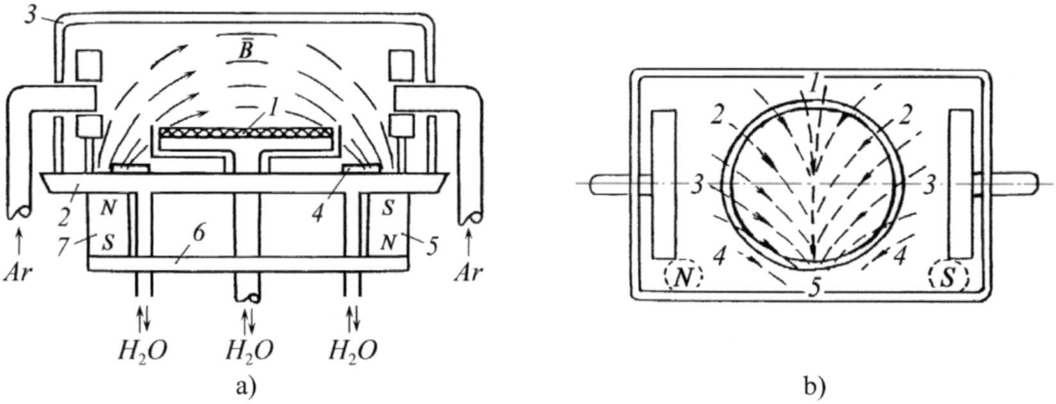

Figure 1. Design scheme of HF discharge reactor of magnetron type (a), where 1 is a cooled round cathode target; 2 is a cooled body-electrode; 3 is a removable electrode reactor; 4 is a sample substrate; 5, 7 are permanent magnets; 6 is ferromagnetic plate; and scheme of location of the samples (b), where 1–5 are location of the samples.

HF discharge is ignited between two cooled electrodes 1 and 3 during the inflow of working gas argon according to technological cycle after pumping of residual gases from the reaction chamber volume. HF electrodes 1 and 3 in the gap are exposed to the bombardment by electrons and ions of plasma, which leads to cleaning inner surface of reactor and substrates 4. Then on the electrode 1 the PTFE target begins to dissipate during bombardment by ions of HF plasma discharge that generates working fluorine gas atmosphere containing molecular fragments of fluoropolymer. Molecular polymer particles in plasma HF discharge is additionally subjected to plasma chemical reactions in a gas atmosphere; chemical connections with the active surface centers are formed and they deposited on the substrate 4.

Applied inhomogeneous magnetic field *B* has normal and tangential components, which create uneven etching on the target surface due to its bombardment by oscillating electrons and ions of plasma in a magnetic field (Figure 1a). The impact of charged plasma particles on the growing fluoropolymer film essentially depends on the substrate's location, which is determined by the tangential component of magnetic field *B* on the perimeter of round target 1. So the value *B* has a minimum value in the region of samples 1 and 2 (Figure 1b), which are away from poles of the magnets 5 and 7 (Figure 1a) and it has a maximum value in the region of samples 4 and 5 (Figure 1b) between poles 5 and 7 (Figure 1a). It is possible to control the density of electron plasma on the growing film of fluoropolymer on the substrate by changing the magnetic field *B*, which allows to significantly expanding the technological range of operating pressures and possibilities of regulating the concentration of the active gas atmosphere.

3. Results and Discussion

3.1. HF Discharge Reactor

Experimental dependences of kinetics of the deposition process of fluoropolymer coatings from active gas phase in the cathode HF discharge reactor without the magnetic system were determined.

Figure 2 shows dependences of fluoropolymer coating growth rate *V* from pressure of vapors *P* in the vacuum chamber (a) and substrate temperature T_s (b) with the plasma current density of 0.35

[A/m^2]. The curves 1–5 (Figure 2a) correspond to the substrate temperatures T_s of 20, 30, 50, 70 and 90 [°C]; the curves 1–4 (Figure 2b) correspond to vapors pressure P of 0.3; 0.4; 0.5 and 0.6 [*Pa*] in the vacuum chamber.

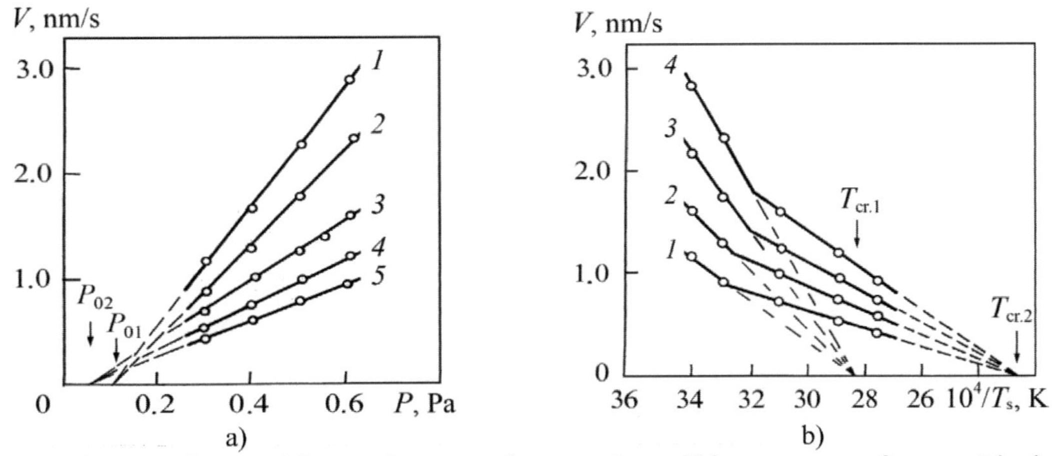

Figure 2. Dependences of fluoropolymer coating growth rate V from pressure of vapors P in the vacuum chamber (a) and substrate temperature T_s (b) with the plasma current density of 0.35 A/m^2.

Rate of V of coating growth significantly decreases at increasing vacuum and substrate temperature (Figure 2). The vapors pressure P in the chamber at decomposition by plasma of a block polymer PTFE has a threshold critical value of ~ 0.07...0.1 [Pa]; below this value the process of deposition of a layer of fluoropolymer on the substrate does not occur.

Dependence of fluoropolymer film V on temperature finds out the two regions corresponding to two types of the kinetics of coatings deposition. It allows determining the optimal temperature of resistant polymerization of molecular components of gas atmosphere, which is in the region of the critical temperature $T_{sr.1}$ and $T_{sr.2}$ between of 78°C and lower of 165°C. At a defined temperature polymer V exceeds the diffusion rate of deposited molecules into solid phase of polymer, where the propagation of reaction occurs. Reduction of V with increasing temperature is similar to the process of depolymerization and reduction of the fluoropolymer molecular weight. It plays an important role on impact growing layer of electrons and ion plasma flow [8, 9] and the high-frequency electromagnetic fields [10]. Here the impact on the growing layer of electron flow and plasma ions [8, 9] and the high-frequency electromagnetic fields [10] plays an important role. So polymer with the maximum value of molecular weight (870 000...1 080 000) is formed in the film in the temperature range of 78...150°C [11]. The most optimal molecular structure of the fluoropolymer film with a molecular weight is formed at the substrate temperature in the range of 95...127°C; it coincides with the results of the experiment at the PTFE synthesis from monomer [8]. The morphology of the films significantly depends on the synthesis temperature. So on cold substrate a fluoropolymer structure is amorphous and homogeneous with crystallinity not more of 1.5%. Crystallinity of the films increases up to values of 15...18% at increasing temperature; the surface morphology is expressed by some relief.

3.2. HF Discharge Reactor with magnetic field

According to the considered design of HF discharge reactor with heterogeneous magnetic field the impact of the plasma flow on the growing polymer film will be different. So this impact will be minimal in region 1 of the reactor (Figure 1b); it will be maximum in the region of strong magnetic field 4 and 5. Fluoropolymer film in region 1, where there is practically no influence of the plasma on the growing film, has a molecular structure similar to the original PTFE with crystallinity values of 25...35%.

Fluoropolymer structure contains a wide range of components (short, medium and long) of the molecular chains of tetrafluoroethylene and also there is a large part of radiation-grafted chains with the group of oxides and radicals in the region of the substrates 4 and 5 (Figure 1b), where the intense ionization of the growing polymer film occurs. The crystallinity of such films does not exceed of 3...5%. The morphology of last specimens has a more homogenous surface due to the high degree of cross-linking of molecules during exposure of the plasma discharge.

Figure 3 shows dependences of mass growth (of thickness L) (Figure 3a) and relative value of growth speed $V(t)/V_0$ (Figure 3b) on time t of fluoropolymer films from the gas phase on silicon substrate at temperature T_S = 23 °C. Curves were obtained at the density of discharge current on the surface of the growing film, where 1 – 0.05; 2 – 0.3; 3 – 0.63; 4 – 0.95 and 5 – 1.4 [A/m^2]. The V_0 values for the curves 1, 2, 3, 4 and 5 are 0.9, 1.36, 1.98, 2.41 and 2.8 [mg/m^{2}s] respectively.

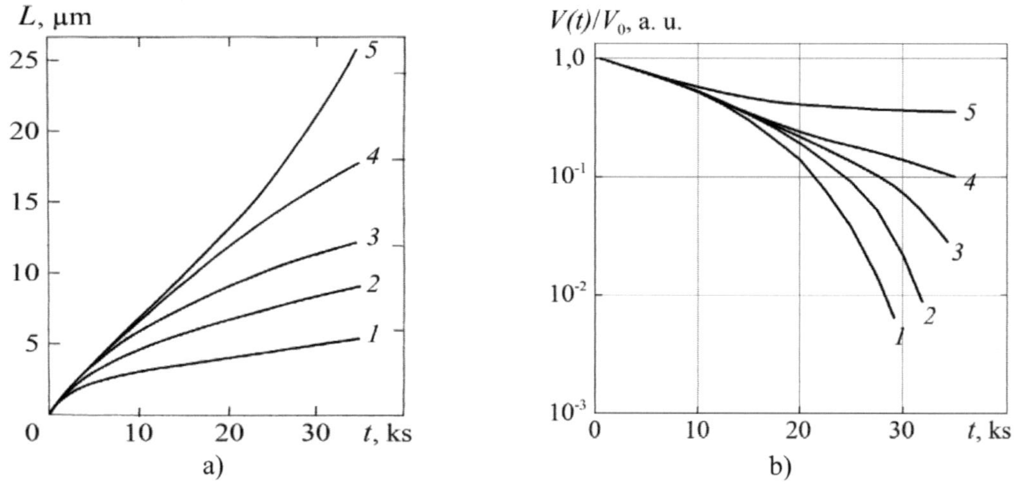

Figure 3. Dependences of mass growth (of thickness L) (a) and relative value of growth speed $V(t)/V_0$ (b) on time t of fluoropolymer films from the gas phase on silicon substrate at temperature T_S = 23 °C.

Figure 3 shows that the kinetics of fluoropolymer films growth in HF discharge reactor (Figure 1a) is strongly dependent from the location of the specimens 1–5 in the reactor (Figure 1b). Deposition process of fluoropolymer for specimens 1 and 2 (Figure 1b) has the first order; it becomes zero order at the increasing flow of electronic processing of growing film (from region 2 to specimens 4 and 5), i.e. radiation-induced polymerization of molecular fragments from the gas atmosphere on the substrate is dominated. The deposition rate of the coating $V(t)/V_0$ from the gas atmosphere is maximum at the initial stages; it is determined by the constancy of the electron-stimulated polymerization processes at increasing layer thickness L to some stationary value. Research of various modes of HF discharge reactor are showed that the growth rate of the film $V(t)$ depends on the value of the tangential component of rms value of constant magnetic field B between the capacitive electrodes 1 and 3 (Figure 1a), i.e. it depends on the intensity of the electronic processing of the growing polymer layer. Here radiation polymerization processes of adsorbed molecules on the surface of the substrate are played by the decisive role.

Analysis of kinetic dependences of specimens in region 1 and 2 on the reactor (Figure 1b), where negligible influence of active component of the gas phase on the formed polymer layer is, has indicated that the polymerization reaction has a first order (Figure 3, curves 1-3). Order of the growth kinetics changes from first to zero order (Figure 3, curves 4 and 5) at the coating thickness of 3...5 μm in the region of specimens 4 and 5 (Figure 1b), where the polymerization reaction proceeds rapidly. For the intermediate region 3 this transition occurs only at the final stage of the coating formation. This phenomenon in the film deposition process can be explained by the fact that the active centers of substrate are the deciding factor of the polymer film growth in the initial period of the layer condensation

Films are formed according to the mechanism of the adsorption-type polymerization at the absence of an electron-ion bombardment, when the neutral particles and macromolecules of products of plasma chemical reactions are deposited on the surface. For the case of intense plasma exposure on growing layer of fluoropolymer (Figure 3, curve 5) radiation-stimulating processes of transverse molecular linkings with the formation of cluster structures are dominated [8, 9]. Processes of temperature increasing of growing polymer layer due to the impact of high-frequency field [10-14] and the plasma particles of HF discharge [15, 16] also have an important role. Moreover, the short molecular -C-C- structures of main chain dominate in the polymer; it substantially impairs their electrical properties, but increases the adhesion and cohesion characteristics of the coating.

5. Conclusion

Technological HF discharge reactor with oscillating electrons in magnetic field, which allows significantly extending the technological range of operating pressures and possibilities of regulating the concentration of the active gas atmosphere, is proposed. Influence of various technological factors, which occur during the formation of polymer coatings from active gas atmosphere in HF discharge, is considered. The most effective directions of intensifying the process of deposition of coatings in HF discharge are investigated; the fields of optimal mode of the technological process of formation of the fluoropolymer coating are defined. The results of the conducted experimental studies are showed that polymerization process, the molecular and macroscopic structure of fluoropolymer films are mostly determined by the temperature mode of deposition of the molecular flow and intensity of exposure of the active gas atmosphere.

Acknowledgments

This work was supported by the Ministry of Education and Science of the Russian Federation (No 3.8074.2017/BCh, basic part of the State Task).

References

[1] Yasuda H 1985 *Plasma Polymerization* (Orlando, FL: Academic Press Inc.)
[2] Rogachev A A 2014 *Physic and chemistry of polymer coatings deposited from active gas phase* (Moscow: Nauchnii mir) (in Russian)
[3] Luchnikov P A *et al* 2016 *IOP Conf. Ser.: Mater. Sci. Eng.* **110(1)** 012081 DOI: 10.1088/1757-899X/110/1/012081
[4] Luchnikov P A 2015 *Russ Phys J* **57(9)** 1239–1245 DOI: 10.1007/s11182-015-0369-0
[5] Ragachev A A *et al* 2015 *Appl. Surf. Sci.* **351(1)** 811–818 DOI: 10.1016/j.apsusc.2015.06.008
[6] Aleutdinov A D *et al* 2015 *IOP Conf. Ser.: Mater. Sci. Eng.* **81(1)** 012069 DOI: 10.1088/1757-899X/81/1/012069
[7] Yarmolenko M A *et al* 2016 *Micro- and Nanocomposite Polymer Coatings Deposited from Active Gas Phase* (Moscow: Radiotekhnika) (in Russian)
[8] Luchnikov P A *et al* 2017 *IOP Conf. Ser.: Mater. Sci. Eng.* **168(1)** 012092 DOI:10.1088/1757-899X/168/1/012092
[9] Piliptsou D G *et al* 2017 *IOP Conf. Ser.: Mater. Sci. Eng.* **168(1)** 012103 DOI:10.1088/1757-899X/168/1/012103
[10] Nefedov V I *et al* 2017 *IOP Conf. Ser.: Mater. Sci. Eng.* **168(1)** 012098 DOI:10.1088/1757-899X/168/1/012098
[11] Enslin S E *et al* 1986 *South African Journal of Chemistry* **36(1)** 23–26
[12] Surzhikova O A 2015 *IOP Conf. Ser.: Mater. Sci. Eng.* **81(1)** 012098 DOI: 10.1088/1757-899X/81/1/012098
[13] Yakunin M A *and* Yurchenko A V 2015 *Technical Physics* **60(1)** DOI: 10.1134/S1063784215010272
[14] Makarov V V *et al* 2017 *IOP Conf. Ser.: Mater. Sci. Eng.* **168(1)** 012005 DOI: 10.1088/1757-899X/168/1/012005
[15] Pevtsov E Ph *et al* 2017 *IOP Conf. Ser.: Mater. Sci. Eng.* **168(1)** 012095 DOI:10.1088/1757-899X/168/1/012095
[16] Luchnikov P A *et al* 2017 *IOP Conf. Ser.: Mater. Sci. Eng.* **168(1)** 012097 DOI:10.1088/1757-899X/168/1/012097

IOP Conf. Series: Materials Science and Engineering **189** (2017) 012015 doi:10.1088/1757-899X/189/1/012015

Algorithm and Software for Landslide Slopes Stability Estimation with Online Very Low Frequency Monitoring

V F Gordeev[1], M M Kabanov[2], and S N Kapustin[2]

[1]Senior Researcher, Institute of Monitoring of Climatic and Ecological Systems, Siberian Branch of the Russian Academy of Sciences, Tomsk, Russia
[2]Junior Researcher, Institute of Monitoring of Climatic and Ecological Systems, Siberian Branch of the Russian Academy of Sciences, Tomsk, Russia

E-mail: gordeev@imces.ru

Abstract. In addition to preliminary surveying, landslide slopes stability estimation problems require online real-time monitoring alerting about potential emergencies. Very low frequency monitoring data provided by geodynamic processes automated control system provides a solution to that problem. Authors describe the software and algorithms implemented for that system, make conclusions on the efficiency of applied solutions and propose options for the further development of online very low frequency monitoring system.

1. Introduction

Landslide stability estimation is heavily used in engineering and geodetic surveying. Such methods are mostly based on the quantitative slope stability estimations prior to design and construction works at a respective site [1]. However, lately a new research direction is developing rapidly, focusing on online monitoring, estimation and prediction of soil engineering stability. In such approach Earth's natural pulsed electromagnetic field (ENPEMF) recorded data could be an essential information source [2–4].

To measure those fields Multichannel geophysical recorders "MGR-01" were designed and built at IMCES SB RAS [5]. "MGR-01" records pulse signals by electric and magnetic components of the field, and for the magnetic component measurements are performed in a narrow band on very low frequencies (VLF) using two mutually perpendicular directions for signal reception. "MGR-01" recorders are certified (certificate No 24181), registered in a state registry of measurement devices and legal to use in Russian Federation. Based on those devices, fixed networks accumulating the real-time data were created [6]. In addition, two new types of device were introduced to the networks lately – MTZ recorder having three recording channel instead of two and GR recorder with a single recording channel.

To aggregate all the accumulating data and to allow various user groups to access that data, the software was developed as a part of a geodynamic processes automated control system (GPACS).

2. GPACS software general structure

The general structure of GPACS software is outlined on Figure 1.

Software was implemented on the platform of Java 2 EE + JBoss Application Server, using MySQL RDBMS as a data repository.

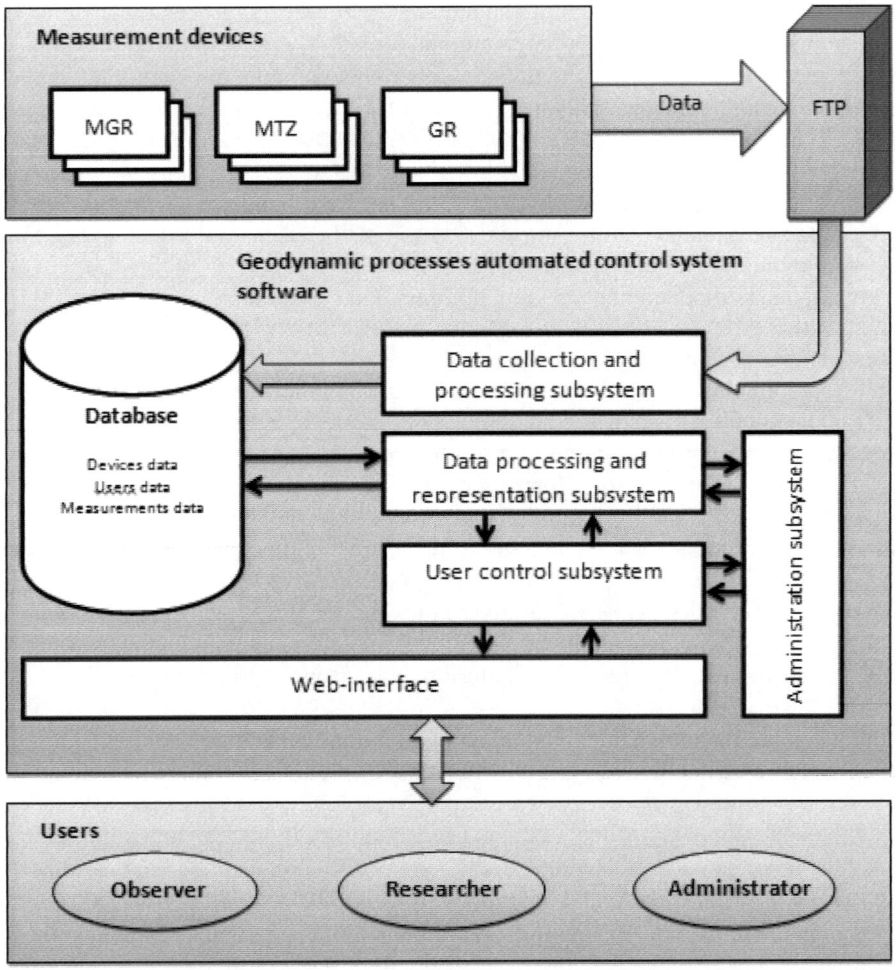

Figure 1. General structure of GPACS software.

The following core subsystems can be identified in the software:

1. Data collection and processing subsystem. This subsystem performs the input of the data coming from the devices into the database. More detailed description follows in section "Data collection from measuring devices in GAPC".
2. Administration subsystem. That functionality is designed to manage device service data, user groups and system overall settings.
3. User control subsystem. This subsystem performs access differentiation for separate user groups varying in functionality, data visibility and interface layout.
4. Data processing and representation subsystem. Major functionality part of the system, providing features to filter data for a user, processing data to detect dangerous geological processes, providing features to either review the data in web-interface or export it in user-specified format.
5. Web-interface. A thin client working in any web browser, responsible for user interaction with the system.

3. Data collection from measuring devices in GAPC

Recorders data is output in fixed data interval and accumulated on ftp server as files with the naming pattern of XXXXX_DDDDDD_TTTTTT.mgr, where XXXXXX is a recorder descriptor, DDDDD is measurement date, TTTTTT is output time. Currently recorders export their data once in four hours.

IOP Conf. Series: Materials Science and Engineering **189** (2017) 012015 doi:10.1088/1757-899X/189/1/012015

System automatically imports measurement data from ftp servers aggregating it into the database. During the import the following data operations are performed:

1. For each recorder a control channel E value is checked and the data considered invalid is excluded from the representation and analysis. That data is however stored in the database to observe a respective device statistics – if invalid data threshold is exceeded for a recorder – the hardware check is scheduled.
2. Hourly average values are calculated and stored separately in the database. While it looks like a theoretical violation of database design normal form, it still makes a practical sense, because those particular average values are heavily used in further analysis and it is more efficient to duplicate information rather than sacrifice computational efficiency later on.

4. Online monitoring software

One of the key requirements for the system design was a simple data access in any circumstances. Therefore, it was a logical choice to use a thin client architecture, where any workstation with an internet connection and web browser could be used to access the system. Software allows to work via web-interface with either a single station (recorder) or a set of stations. The following functionality is available:

1. Output measurement data for a selected period of time. Both raw and hourly average data can be viewed.
2. Export measurement data to tab-separated values file for the further analysis in any third-party software.
3. Review measurement data plots for a station or a group of stations – either raw or hourly average.
4. Review the plots composed by the stability estimation algorithm. Described in more detail in the section "landslide slope stability estimation algorithm".

In addition, there is a number of service features in the system, including administrative interface to set up graphic representation, user groups, station groups, ability to review the device statistics and so forth.

Overall functionality of the system makes it useful for the following three user categories:

1. Observer, performing online monitoring of landslide slope stability. With the help of the algorithm he can make a decision on the necessity of on the ground inspection.
2. Researcher, analyzing temporal sequences of accumulated data to suggest new decision strategies and distinguish possible underlying tendencies.
3. Devices network administrator, performing maintenance control of the recorders and managing the system usability.

5. Landslide slope stability estimation algorithm

Landslide slope stability estimation algorithm based on ENPEMF uses two criteria. The first criterion indicates difference in the intensity of the measurements between a set of recorders located at the stress-strained rocks state control sites and a reference recorder. Reference recorder is placed no farther than 25 km (VLF wave length) from control recorders at a location, which is not a subject to geodynamic processes. Another criterion estimates a degree of similarity in different recorders readings. Excess in the intensity of ENPEMF pulse flow in each measurement point relative to reference station is calculated using Equations:

$$K_{(N-S)} = \frac{(N1_{T(i)} - N1_{T(r)})}{N1_{T(r)}}, \qquad K_{(W-E)} = \frac{(N2_{T(i)} - N2_{T(r)})}{N2_{T(r)}},$$

where $N1_{T(i)}$ is ENEMPF pulse flow intensity in north-south direction on i-th measurement station; $N2_{T(i)}$ is ENEMPF pulse flow intensity in west-east direction on i-th measurement station; $N1_{T(r)}$, $N2_{T(r)}$ is ENEMPF pulse flow intensity in north-south and west-east direction on reference measurement station.

If coefficients $K<0$, then we consider that the mechanical compression strains emerge in a respective direction, either north-south or west-east. If $K>0$, tension strains emerge. The greater the value of K, the higher are strains. That parameter is one of the most essential criteria for ground conditions geodynamical estimation.

The second parameter for stress-strained ground state is the estimation of correlation between the forms of intraday ENPEMF intensity variations of reference station and i-th control station. To calculate correlation coefficient of two samples Spearman's Equation is used [7]:

$$r = 1 - \frac{6}{n^3 - n} \sum_{i=1}^{n} (x_i - y_i)^2 ,$$

where n is sample size (number of measurements, used for the calculation); x_i is first sample i-th element's rank in that sample (e. g. sample is {5, 8, 4, 6}, then $x_1 = 2$, $x_2 = 4$, $x_3 = 1$, $x_4 = 3$); y_i is second sample i-th element's rank in that sample.

Further on, integrated coefficient for landslide slope geodynamical state stability is calculated using the Equation:

$$R_a = \frac{k + (1 - r)}{2} ,$$

where R_a is aggregated criteria for the residual stability of a particular location, which value is determined empirically.

It should be considered dangerous for an industrial site if integrated coefficient exceeds 150% for three days or more.

Example of integrated coefficient plot can be seen in the Figure 2.

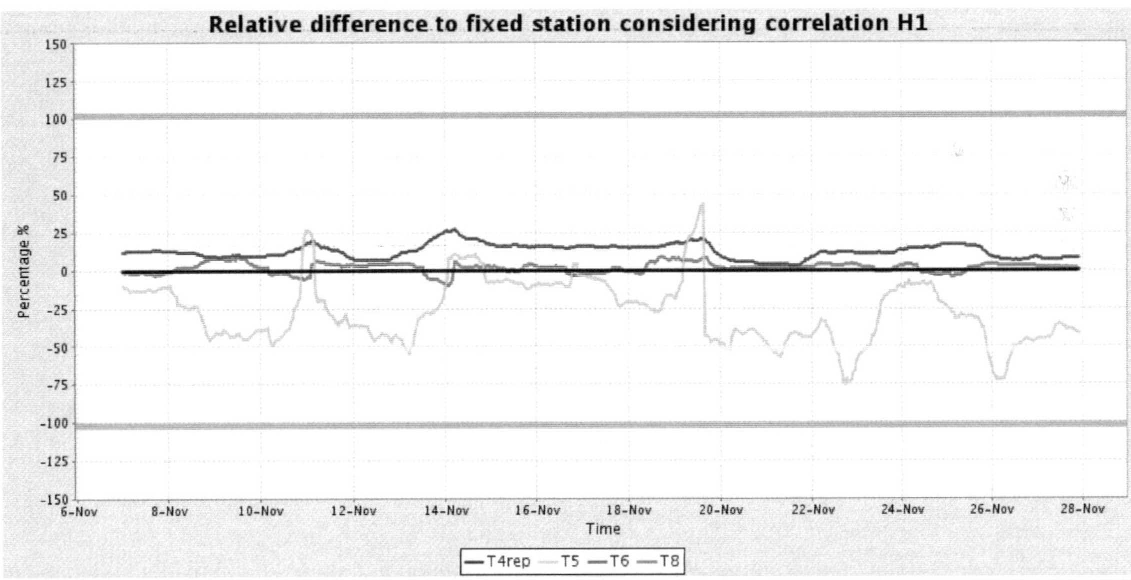

Figure 2. An example of integrated coefficient used for a set of four stations. Reference station is T10.

6. Summary

Suggested application for ENPEMF signal processing of the spatially distributed network of stations provided accurate, reproducible results, reflecting geodynamical processes activity with a very high confidence. The system allows to determine landslide hyperactivity and hypoactivity zones, tension and compression strains zones, spatial orientation of stress for a slope structure in real-time.

During the system operating time the anomalies in ENPEMF pulse flow were observed over 10 times on different section of the slope. Anomalies manifested as distortions in diurnal trajectory of

ENPEMF intensity for several days. In those days there also was a substantial difference in the reading of the stations located on different sections of the slope.

As a further development of this work it seems logical to enhance the system autonomy and to shift the focus in landslide danger decision making process from a human operator to a fully automated algorithm. That would inevitably require a higher quality of critical situation automated recognition. Random processes change point detection algorithms looks a promising direction of development [8]. Within that approach it is proved possible to construct a procedure automatically detecting a change in the mean or variance of a random process and moreover determine the parameters of such procedure providing optimal ratio between a delay in change detection and a number of false alarms.

References

[1] Fredlund D *and* Krahn J 1977 *Canadian Geotechnical Journal* **14(3)** 429–439 DOI: 10.1139/t77-045

[2] Malyshkov Yu P *and* Malyshkov S Yu 2012 *Eccentric rotation of the earth's core and lithosphere: Origin of deformation waves and their practical application* (Taiwan: Nova Science Publishers) **6** 115–212

[3] Malyshkov Yu P *et al* 2013 *Electromagnetic Fields: Principles, Engineering Applications and Biophysical Effects* (Taiwan: Nova Science Publishers) 1–86

[4] Gordeev V F *et al* 2016 *IOP Conf. Ser.: Earth Environ. Sci.* **48** 012026 DOI: 10.1088/1755-1315/48/1/012026

[5] Shtalin S G *et al* 2012 *Sensors & Systems* **4** 32–37 (in Russian)

[6] Malyshkov S Yu *et al* 2016 *Proc. of Int. Conf. GEO-SIBIR'16* **2** 13–18 (in Russian)

[7] Myers J L *and* Well A D 2003 *Research Design and Statistical Analysis* (U.S.: Lawrence Erlbaum Associates Inc.)

[8] Vorobeichikov S E *and* Kabanova T V 2002 Journal of Communications Technology and Electronics **47(10)** 1198–1203

X-ray Tomographic System Behavior Prediction Based on a Mathematical Model

S S Baus[1], L A Redko[2], and M N Yanushevskaya[2]

[1]Student, National Research Tomsk Polytechnic University, Tomsk, Russia
[2]Associate Professor, National Research Tomsk Polytechnic University, Tomsk, Russia

E-mail: vela2007@bk.ru

Abstract. There appear certain challenges in defining the dependence of the X-ray radiation intensity change in passing through the material (as fixed by the detector) conditioned by various parameters of an X-ray optical system while designing new modifications of X-ray tomographs. At present, this problem is experimentally solved by selection of voltage corresponding parameter values on an X-ray tube with thickness and type of the studied material considered. To reduce the design time and complexity, a mathematical model of parameter behavior is required to characterize the X-ray optical system in the major working range of values. The present paper investigates the X-ray optical system behavior using methods of mathematical statistics. A regression model has been obtained which matches the change of the X-ray intensity value to the intensity in the X-ray tube. The research has defined the further study direction of X -ray optical system parameters.

1. Introduction

X-ray tomographs are regarded the most widespread systems for industrial non-destructive testing (NDT). X-ray tomographs are classified into several types according to the studied value sizes, their technical characteristics, the field of application, etc [1]. Currently, microtomographs are considered the most popular direction of NDT system development [2]. These systems allow investigating internal structure of objects with micron sizes and are characterized by voltage on the X-ray tube in the range from 10 to 160 [kV] and a small size of the focal spot (from 1 to 10 microns). As betatrons belong to another X-ray system class due to an alternative implementation principle and their technical characteristics, they are not considered in this work.

An X-ray optical system for NDT consists of the following four elements: an X-ray tube, a detector, a worktable positioning system and the test material [3]. The X-ray tube and detectors (Hammatsu, Japan) produced in Tomsk have been used in microtomographs. These elements fully comply with modern advanced requirements in the field of energy saving, functionality, measurement accuracy and an X-ray beam [3]. These tomographs compete with the tomographs produced by such leading companies as Matrix, SkyScan, and Feinfocus.

The X-ray tomograph system design as a fundamental stage of the process involves basic technical system parameters calculations according to the specification, fundamentally new technical solution development as well as the system layout.

As X-ray system development and design is of experimental nature, the equipment and system parameters are selected according to the theoretical formulas of the X-ray intensity change in passing

IOP Conf. Series: Materials Science and Engineering **189** (2017) 012016 doi:10.1088/1757-899X/189/1/012016

through material and intuitive component selection conforming to these parameters. These calculations are considered labor-consuming and mostly characterize a situational (static) condition of an X-ray optical system within specifically chosen factor values and do not allow complete system behavior (change of output parameter) estimating at factor value changing [4].

2. Problem statement

To avoid such complexities, a new method of determining mathematical relationships has been required to enable the process simulation and system behavior analysis, to reduce time costs and to improve the developed system quality and accuracy [5]. To achieve this goal, the tools and features of a complete factorial experiment have been applied. A complete factorial experiment allows to vary all factors simultaneously and to obtain a regression model of the studied system [1]. In addition, it enables the search of optimum conditions of functioning of the system based on the adequate models.

3. Research questions

The following tasks have been set and solved to achieve the abovementioned goals:

• to define the X-ray optical system layout;
• to determine the system optimal values;
• to conduct a complete factorial experiment 2^3;
• to conduct a regression analysis;
• to obtain a mathematical relationship for predicting the process of X-ray intensity changing for the basic operation modes and studied materials;
• to evaluate the obtained results and their applicability in practice.

This value of response is considered significant as it characterizes the quality of shadow images obtaining and 3D model building, uniformity and quality of the X-ray tube emitting beam, the material parameters and its uniformity, etc.

The experiment has covered three basic factors of the process:

1. the voltage at the X-ray tube;
2. the thickness of the illuminated material;
3. the type of material.

The basic factor levels have been formed; they characterize the main modes of the device operation. The variation intervals have been formed on the basis of the real range of factor acceptable values. The tables of the parameter actual values and their intervals have been filled in according to the experiment plan X.

Table 1 demonstrates the two materials (titanium and iron) selected for the experiment.

Table 1. Factor levels and variation intervals.

Factor levels	Designation	U, [kV] $\tilde{x}_1$	Thickness, [mm] $\tilde{x}_2$	Type of material $\tilde{x}_3$
Lower	-1	100	1	iron
Upper	+1	140	2	titanium

Iron is one of the most frequent materials applied in various industrial detail production and parts in mechanical engineering, space engineering as well as in the radio-electronic equipment. Titan has a broad application in the aircraft equipment, rocket production, instrument making, in chemical industry and shipbuilding [6].

4. Research methods

The work aims are to improve the process of new X-ray optical system design; to reduce time and labor input due to forecasting of X-ray optical system behavior based on a mathematical model.

The applied cover methods and tools for X-ray system design, theoretical calculation methods of X-ray radiation intensity changes, the two-level, complete factorial experiment for three factors, namely the 2^3 design, dispersive and regression analyses are used as the research methods.

First, the X-ray microtomograph was plugged into the electrical network. After 2-3 minute waring up, the check run of X-ray radiation pulse without material was produced.

The value of the X-ray intensity was measured. The experiment could be further conducted if the value observed was comparable to the device technical documentation or the X-ray intensity deviation cause had been eliminated. With the system working properly, the material translucence started. First, the material was placed on the device worktable and the lid was closed for safety reasons. Then, the voltage value was set on the X-ray tube and after the signal of material transmission the result presented on the display device was recorded. The studied material, its size and the X-ray tube voltage value were to be chosen according the experiment plan.

The experiment aimed to study the process of changing X-ray intensity obtained in the X-ray microtomograph in passing through the material (iron and titanium) of two sizes 1 and 2 mm respectively. Factor alternating was carried out for obtaining a set of parameter combination values listed in the rows of the plan table. The experience was repeated 3 times at each point of the experiment plan. Table 2 shows the test results (value of X-ray intensity).

Table 2. Experimental results.

No Exp.	Factors in the natural scale			Factors in the dimensionless coordinate system				Output parameter		
	Z_1 [kV]	Z_2 [cm]	Z_3	X_0	X_1	X_2	X_3	Y_1 [keV]	Y_2 [keV]	Y_3 [keV]
1	100	0.1	1	+1	-1	-1	-1	30.4	30	29.6
2	140	0.1	1	+1	+1	-1	-1	65.6	66.4	63.8
3	100	0.2	1	+1	-1	+1	-1	14	13	13.4
4	140	0.2	1	+1	+1	+1	-1	46.31	46.2	45.7
5	100	0.1	2	+1	-1	-1	+1	49.6	48.2	48.4
6	140	0.1	2	+1	+1	-1	+1	79.8	80.6	80.7
7	100	0.2	2	+1	-1	+1	+1	37	37.1	36.8
8	140	0.2	2	+1	+1	+1	+1	69.8	70	68.6

5. Findings

Multiple measurements under identical conditions were carried out to reduce experimental errors. Dispersion homogeneity testing was performed by using the Cochran's criterion.

As a result, the regression equations were received:

- a linear equation, which considered only the coefficient factors, $Y=48.8+16.5\ x_1 - 7.3\ x_2+10.1\ x_3$;
- a nonlinear one, which regarded the coefficients of individual factors and their interaction with each other, $Y=48.8+16.5\ x -7.3\ x_2+10.1\ x_3 - 0.47\ x_1\ x_3 +1.65\ x_2\ x_3+0.458\ x_1\ x_2\ x_3$.

The equations presented the research results: the larger the absolute value of the regression coefficient b_i, the greater its influence on the optimization criterion in a specified range factor values.

The model adequacy was then checked by the Fisher criterion which led to the conclusion that the two obtained equations were not adequate. The model poorly described the experimental data in the specified range factor values. If a linear model was inadequate, then that could not be approximated by a plane surface response. Some other factors affecting the model have not been considered in the experiment described. There is a suggestion that the materials selected for the experiment possess different physical properties. The parameter value dispersion characterizing the linear attenuation coefficient of X-rays in the material is equal to 0.6 cm^{-1} for iron up to 1.9 cm^{-1} for titanium respectively when the energy is about 0.7 MeV. Furthermore, titanium has a more complicated three-dimensional internal structure and an X-ray attenuation dependence which differs from that of iron.

IOP Conf. Series: Materials Science and Engineering **189** (2017) 012016 doi:10.1088/1757-899X/189/1/012016

The decisions generally taken to build up an adequate model might be the following: to change intervals of varying factors, to shift the focus of the experiment plan, to extend the experimental design.

Among these, the most common method is to adjust the range of values of varying factors. This method is based on carrying out a new series of experiments [7]. However, this solution does not seem to be appropriate in this situation, because the intervals of variation have been defined on the experiment planning stage and include important materials in modern industry and meet the basic operating conditions of an X-ray system. Changing this interval may result in changing the range of values, which is of the greatest interest from the point of view of practical applications, including the construction of a model to predict the results. Altering the range of values of varying factors might turn possible, if materials with similar physical properties are selected [8].

The study continued as univariate regression analysis. Then, the impact of the effect of an X-ray tube voltage value on the value of the output intensity after passing through the material under investigation is considered.

An additional experiment was conducted under the following conditions. The values of thickness and type of the material were fixed, as well as the of X-ray tube power. The selected material was iron, as one of the most frequently encountered in practice. The investigated iron plate had a thickness of 2 mm. the Power of the X-ray tube was kept the same, 0.7 [MeV].

The following variables were identified for analysis:

1. The output power intensity after passing through the material under investigation, [keV] – Y;
2. The voltage on the X-ray tube, [kV] – X.

Regression analysis is usually used in order to determine the correlation ratio between indicators of the voltage and the output intensity of the evaluation factors that have the greatest impact on output intensity. It also used to establish the degree of the output intensity on the stress influences and to select the type and form of unknown causal relationships, as well as to define the calculated values of the output intensity (regression function). Regression analysis allows predicting the value of the output intensity for better construction of a 3D shadow image when working with the instrument. The most important stage of building a regression model (regression equation) is the selection of the mathematical function form that most accurately defines the existing relations between the analyzed features [9–11]. In this case, intensity has been assigned as the depended attribute of the magnitude, and stress as the independent attribute respectively.

The raw data analysis allows suggesting that the pair linear correlation equation is represented by the function:

$$y=bx_0+bx_1.$$

Figure 1 shows a dispersion diagram. The correlation has a positive trend.

Performing the corresponding calculations in Excel the following results have been obtained. The sums of squares calculated in Excel stand for SSR=3890.653; SSE=107.728; SST=3998.3811. R-squared (coefficient of determination) is a utility assessment of the regression equation and estimates the proportion of the Y variation of the variable, which is explained by the independent variable in the regression model X. This value is equal to 0.973. It describes that 97.3% of the variation is due to the intensity of the output voltage value and only 2.7% by other factors. Large values tending to 1, indicate that X can be used to explain the behavior of Y.

The standard error is equal to 3.6696. This shows that 95% of the values may fall within ±3.6696 from the regression line.

As a result, a linear regression equation has been obtained: $Y= - 55.9916+0.6867\cdot x$. This equation can be applied to the selected range of operating systems values. The positive slope of the linear regression indicates that an increase in voltage in one unit leads to the rise value of the output intensity at 0.6867. For example, for a voltage of 120 [kV] the output intensity value is equal to 26.41 [keV].

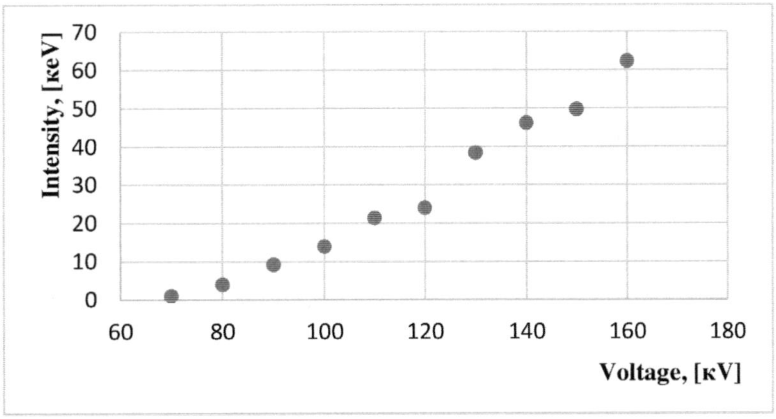

Figure 1. Dispersion diagram.

6. Conclusion

The complete factorial experiment has proved that the resulting linear model is inadequate. This means that the model poorly describes the experimental data in the specified range factor. To use these equations in order to predict the intensity value changes as it passes through the material under investigation is unfeasible. The solution to the problem is to continue the research on the model of a two-factor experiment, excluding the factor of the material. Thus, the principle factors are considered the X-ray tube voltage and the thickness of the material as the. X-ray optical system developers might get interested in the experiment results.

Due to the obtained result there is a need to continue research in order to study this process and provide useful information for the development and design of new versions of X-ray scanners. For this purpose some additional experiments have been conducted. Univariate regression analysis has been carried out where the response of the system as a complete factorial experiment has been the intensity value after passing through the thickness of the material, and the an X-ray tube voltage has been taken as a factor. All the remaining parameters of the system are a group of «experimental conditions».

Due to the univariate regression analysis, the form of the linear regression equation as been determined, it has been proved that this model could be applied in practice, and there is a correlation between variables X and Y. The practical application of the linear regression equation facilitates predicting the value of the output intensity fast and easily according to the selected voltage on the X-ray tube for this range of values. The system can be described most accurately and completely within the prescribed framework with the dependence derived from practical data. This fact confirms the mathematical relationship from the available theoretical calculation tools. This model has a practical significance, since it allows obtaining better shadow images to construct a 3D image, and facilitates the work of engineers and constructs to develop new versions of X-ray systems.

References

[1] Landisa E N *and* Keane D T 2010 *Materials Characterization* **61(12)** 1305–1316 DOI: 10.1016/j.matchar.2010.09.012

[2] Dunsmuir J *et al* 2006 *Powder Diffraction* **21(2)** 125–131 DOI: 10.1154/1.2204956

[3] Syryamkin V I *et al* 2011 *Russian Journal of Nondestructive Testing* **47(11)** 754–764 DOI: 10.1134/S1061830911110076

[4] Marusina M *and* Kaznacheeva A 2006 *Modern Tomography Types* (St. Petersburg: SPbGU ITMO) (in Russian)

[5] Natterrer F 2001 *The Mathematics of Computerized Tomography* DOI: 10.1137/1.9780898719284

[6] Plotnikova I *et al* 2016 IOP Conf. Series: Materials Science and Engineering 132 012023 DOI: 10.1088/1757-899X/132/1/012023

[7] Adler Ju *et al* 1976 *Planning of Experiment by Search of Optimum Conditions* (Moscow: Nauka) (in Russian)

[8] Box G *and* Wilson K 1951 *J. Roy. Statist. Soc. Ser. B Metho.* **13** 1–45
[9] Masood I *et al* 2014 *Applied Mechanics and Materials* **465-466** 1149–1154 DOI: 10.4028/www.scientific.net/AMM.465-466.1119
[10] Rafidah A *et al* 2014 *Applied Mechanics and Materials* **660** 275–279 DOI: 10.4028/www.scientific.net/AMM.660.275
[11] Redko A *et al* 2015 *EpSBS* **VII** 319–324 DOI: 10.15405/epsbs.2016.02.24

IOP Conf. Series: Materials Science and Engineering **189** (2017) 012017 doi:10.1088/1757-899X/189/1/012017

Statistical Degradation Models for Reliability Analysis in Non-Destructive Testing

E S Chetvertakova[1], and E V Chimitova[2]

[1]PhD Student, Novosibirsk State Technical University, Novosibirsk, Russia
[2]Associate Professor, Novosibirsk State Technical University, Novosibirsk, Russia

E-mail: evgenia.chetvertakova@gmail.com

Abstract. In this paper, we consider the application of the statistical degradation models for reliability analysis in non-destructive testing. Such models enable to estimate the reliability function (the dependence of non-failure probability on time) for the fixed critical level using the information of the degradation paths of tested items. The most widely used models are the gamma and Wiener degradation models, in which the gamma or normal distributions are assumed as the distribution of degradation increments, respectively. Using the computer simulation technique, we have analysed the accuracy of the reliability estimates, obtained for considered models. The number of increments can be enlarged by increasing the sample size (the number of tested items) or by increasing the frequency of measuring degradation. It has been shown, that the sample size has a greater influence on the accuracy of the reliability estimates in comparison with the measuring frequency. Moreover, it has been shown that another important factor, influencing the accuracy of reliability estimation, is the duration of observing degradation process.

1. Introduction

Usually, the estimates of reliability are constructed on the basis of the information about failures. However, in non-destructive testing, the researcher doesn't have any failure time data. Instead of that, there is the information about aging of tested items (the so-called degradation data). In this case, the value of some parameter (index), characterizing the process of degradation (aging) of tested items, is measured. The time moment, when the value of degradation index reaches the critical level, is called the failure time.

Different types of degradation models were considered in Meeker (1998) and Liao and Tseng (2006). The gamma degradation model, which is used when the increments of degradation index have the gamma distribution with shape parameter equal to the difference between the values of trend function at the corresponding moments of time, was considered in Bagdonavičius and Nikulin (2002). The goodness-of-fit test for the gamma degradation model was proposed in Chimitova and Chetvertakova (2014). Tsai et al. (2012) and Chimitova and Chetvertakova (2015) have studied the gamma model with random effects, which describes the unit-to-unit variability. However, this degradation model is suitable for the cases, when the increments of the degradation index are non-negative values. In other cases it is proposed to use the Wiener degradation model, which was considered, for example, by Whitmore (1995) and Tsai et al. (2011). This model includes the normal distribution as the distribution of degradation index increments, where the values of trend function are included in the shift and scale parameters. In addition, one of the advantages of using the gamma and

Wiener degradation models is that reliability estimates can be obtained analytically due to the repeatability of the normal distribution, as well as the gamma degradation model.

In this paper, we propose the usage of the gamma and Wiener degradation models for reliability analysis in nondestructive testing. By means of computer simulation technique, we analyze the accuracy of reliability function estimation. The recommendations for increasing the accuracy by changing the number of items or frequency of measuring the degradation index are given for linear and power trend functions.

2. Gamma Degradation Model

Stochastic process $Z(t)$ characterizing degradation process is referred to as the gamma degradation process, if

- $Z(0)=0$;
- $Z(t)$ is a stochastic process with independent increments;
- increments $\Delta Z(t)=Z(t+\Delta t)-Z(t)$ have the gamma distribution with probability density function

$$f_{Gamma}\left(t;\sigma,\Delta\nu(t)\right)=\left(\frac{t}{\sigma}\right)^{\Delta\nu(t)-1}\frac{e^{-t/\sigma}}{\sigma\cdot\Gamma\left(\Delta\nu(t)\right)},$$

where $\Delta\nu(t)=\nu(t+\Delta t)-\nu(t)$ is the shape parameter and σ is the scale parameter, $\nu(t)$ is a positive increasing function.

If random variables ξ_1 and ξ_2 follow the gamma distribution with scale parameter σ and shape parameters ν_1 and ν_2, correspondingly, then $\xi_1+\xi_2$ follows the gamma distribution with scale parameter σ and shape parameter $\nu_1+\nu_2$. This property explains the fact of using the gamma distribution as a distribution of increments.

Let the mathematical expectation of degradation process $Z(t)$ is

$$M\left(Z(t)\right)=m(t),$$

where $m(t)=\sigma\cdot\nu(t)$ is a trend function of the degradation index.

In this paper, we consider two types of trend functions:

- linear trend function $m(t;\overline{\gamma})=\gamma_0+\gamma_1 t$;
- power trend function $m(t;\overline{\gamma})=\gamma_0 t^{\gamma_1}$.

Taking into account the given assumptions, the stochastic process $Z(t)$ at time moment $t=t_k$ has the gamma distribution with the shape parameter equal to $\dfrac{m(t_k;\overline{\gamma})}{\sigma}$.

The time to failure is the random variable

$$T=\sup\{t:Z(t)<z_0\},$$

where z_0 is the critical value of the degradation path. Then, the reliability function for the gamma degradation model is given by:

$$S(t)=P\{T>t\}=P\{Z(t)<z_0\}=F_{Gamma}\left(z_0;\sigma,\frac{m(t;\overline{\gamma})}{\sigma}\right). \tag{1}$$

Let the realization of degradation path $Z^i(t)$ for the i-th item is denoted as

$$Z^i=\left\{(0,Z_0^i=0),(t_1^i,Z_1^i),\ldots,(t_{k_i}^i,Z_{k_i}^i)\right\},\ i=\overline{1,n},$$

where k_i is the number time moments, in which the degradation index was measured. Then, the sample of independent degradation index increments can be written as:

$$\mathbf{X}_n = \left\{ X_j^i = Z_j^i - Z_{j-1}^i,\ i = \overline{1,n},\ j = \overline{1,k_i} \right\}. \tag{2}$$

Maximum likelihood estimates of parameters σ and $\overline{\gamma}$ are calculated by maximization of the log-likelihood function:

$$\ln L(\mathbf{X}_n) = \sum_{i=1}^{n}\sum_{j=1}^{k_i} \ln f_{Gamma}(X_j^i;\sigma, p_j^i) \to \max_{\sigma,\overline{\gamma}},$$

where $p_j^i = \dfrac{m(t_j^i;\overline{\gamma}) - m(t_{j-1}^i;\overline{\gamma})}{\sigma}$ is the shape parameter of the gamma distribution.

3. Wiener Degradation Model

Stochastic process $Z(t)$ characterizing degradation process is referred to as the Wiener degradation process, if • $Z(0)=0$; • $Z(t)$ is a stochastic process with independent increments; • increments $\Delta Z(t) = Z(t+\Delta t) - Z(t)$ have the normal distribution with probability density function

$$f_{Norm}(t;\theta_1,\theta_2) = \frac{1}{\sqrt{2\pi}\cdot\theta_2} e^{-\frac{1}{2}\left(\frac{t-\theta_1}{\theta_2}\right)^2},$$

where $\theta_1 = m(t+\Delta t) - m(t)$ is the shift parameter and $\theta_2 = \sigma\sqrt{m(t+\Delta t) - m(t)}$ is the scale parameter, $m(t)$ is a positive increasing trend function:

$$M(Z(t)) = m(t).$$

It is well known, that if random variables ξ_1 and ξ_2 follow the normal distribution with shift parameters μ_1 and μ_2 and scale parameters σ_1 and σ_2, correspondingly, then $\xi_1 + \xi_2$ follows the normal distribution with shift parameter $\mu_1 + \mu_2$ and scale parameter $\sqrt{\sigma_1^2 + \sigma_2^2}$.

Then, taking into account the given assumptions, the stochastic process $Z(t)$ at time moment $t = t_k$ has the normal distribution with the shift parameter equal to $m(t_k;\overline{\gamma})$, and the scale parameter equal to $\sigma\sqrt{m(t_k;\overline{\gamma})}$. In this case, the reliability function for the Wiener degradation model is given by:

$$S(t) = P\{T > t\} = P\{Z(t) < z_0\} = \Phi\left(\frac{z_0 - m(t;\overline{\gamma})}{\sigma\sqrt{m(t;\overline{\gamma})}}\right), \tag{3}$$

where $\Phi(\cdot)$ is the standard normal distribution function.

The log-likelihood function for this model can be written as:

$$\ln L(\mathbf{X}_n) = \sum_{i=1}^{n}\sum_{j=1}^{k_i} \ln f_{Norm}\left(X_j^i; m(t_j^i;\overline{\gamma}) - m(t_{j-1}^i;\overline{\gamma}), \sigma\sqrt{m(t_j^i;\overline{\gamma}) - m(t_{j-1}^i;\overline{\gamma})}\right),$$

where $\mathbf{X}_n$ is the sample of independent degradation index increments (2).

4. Investigation of the Reliability Estimates in Degradation Analysis

It is clear, that the accuracy of estimates of reliability functions, obtained on the basis of degradation data, strongly depends on the number of degradation increments. So, it is possible to improve the estimation accuracy by increasing the number of items n and/or the frequency of measuring the degradation index k_i. In this section, we investigate the accuracy of estimation of model parameters and reliability functions (1) and (3) by means of computer simulation technique.

The accuracy of parameters estimation was measured with the value

$$g_N(\theta) = \frac{1}{N}\sum_{i=1}^{N}\left(\hat{\theta}_i - \theta_{true}\right)^2,$$

where N is the number of simulated samples, $\hat{\theta}_i$ is the maximum likelihood estimate of parameter θ by the i-th sample, θ_{true} is the true value of parameter.

The value d_N was used as the measure of accuracy for the reliability function estimates:

$$d_N = \frac{1}{N}\sum_{i=1}^{N}\sup_{t\geq 0}\left|S_0(t)-\hat{S}_i(t)\right|,$$

where $\hat{S}_i(t)$ is the reliability function obtained after parameters estimation by the i-th sample and $S_0(t)$ is the true reliability function.

In Tables 01 and 02 the values of $g_N(\theta)$ and d_N are given for the gamma and Wiener degradation models, correspondingly. We have taken the following true values of parameters:

- $\sigma = 0.5$, $\gamma_0 = 0$ (was not estimated), $\gamma_1 = 0.5$ in the case of linear trend and
- $\sigma = 0.5$, $\gamma_0 = 1$, $\gamma_1 = 1.3$ in the case of power trend function.

The critical degradation value $z_0 = 10$ was taken for both gamma and Wiener models. The number of items $n = 10, 20$ and the number of time moments for each item $k = 10, 20$. The time moments were taken from 0 to 19 with the step $\Delta t = 2$ for $k = 10$ and $\Delta t = 1$ for $k = 20$.

The number of simulations $N = 1000$.

Table 1. The accuracy of estimation of model parameters and reliability function (1)

		Linear trend			Power trend			
n	k	d_N	$g_N(\gamma_1)$	$g_N(\sigma)$	d_N	$g_N(\gamma_0)$	$g_N(\gamma_1)$	$g_N(\sigma)$
10	10	0.3626	0.0063	0.0359	0.2147	0.0199	0.0020	0.0198
20	10	0.2504	0.0030	0.0144	0.1218	0.0144	0.0012	0.0086
10	20	0.318	0.0061	0.0182	0.1693	0.0178	0.0018	0.0091

In Figure 1, the reliability functions $\hat{S}_i(t)$, $i = \overline{1,50}$, obtained for the gamma degradation model with linear trend functions and $n = 20$, $k = 10$ (grey lines), as well as the true reliability function (black line) are presented.

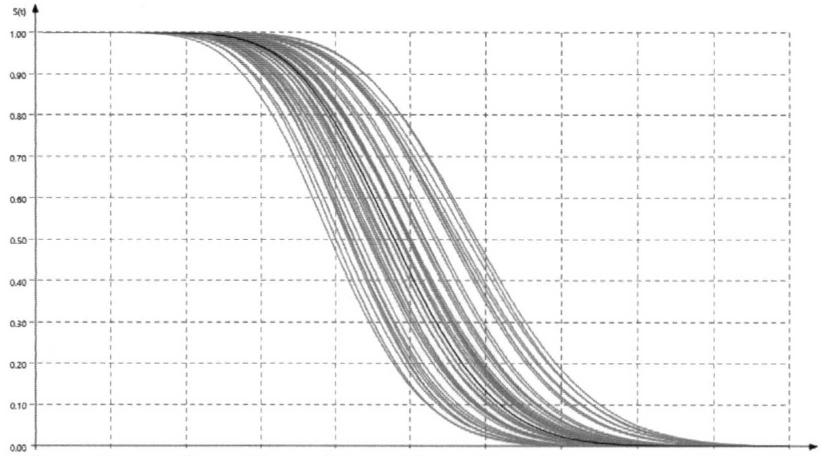

Figure 1. Reliability functions for the gamma degradation model with the linear trend for $n = 20$, $k = 10$, $z_0 = 10$.

Table 2. The accuracy of estimation of model parameters and reliability function (3)

		Linear trend			Power trend			
n	k	d_N	$g_N(\gamma_1)$	$g_N(\sigma)$	d_N	$g_N(\gamma_0)$	$g_N(\gamma_1)$	$g_N(\sigma)$
10	10	0.1161	0.0950	0.0092	0.0898	0.0262	0.0031	0.0070
20	10	0.0888	0.0912	0.0028	0.0673	0.0074	0.0008	0.0035
10	20	0.1124	0.089	0.0038	0.2831	0.0198	0.0022	0.0024

As can be seen from Tables 1 and 2, the estimates, which were obtained in the case of $n = 20, k = 10$, are more accurate for both trend functions and degradation models. It means, that it is more preferable to improve the estimation accuracy by increasing the number of tested items instead of increasing the frequency of measuring.

The accuracy of reliability estimation depends not only on the sample size and measuring frequency. A very important factor is the amount of information about lifetime in observed degradation paths. In Figures 02 and 03, there are the examples of degradation paths with critical levels $z_0 = 10$ and 20, correspondingly. It is obvious, that in the case of $z_0 = 10$ some items have reached the critical level, so the researcher has more information about lifetimes, in contrast to the situation, when the critical level is equal to 20.

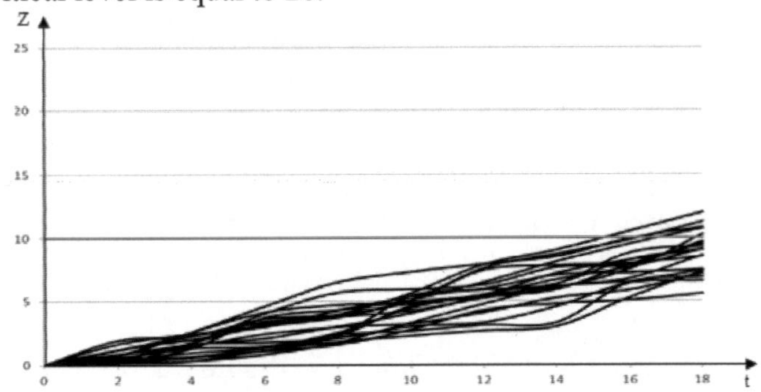

Figure 2. Degradation paths and critical level $z_0 = 10$.

In Figure 4, there are the reliability functions $\hat{S}_i(t)$, $i = \overline{1,50}$, obtained for the same gamma degradation model as in Figure 1, but for $z_0 = 20$. Comparing Figures 1 and 4, one can see, that the reliability estimates are much more accurate in the case of $z_0 = 10$.

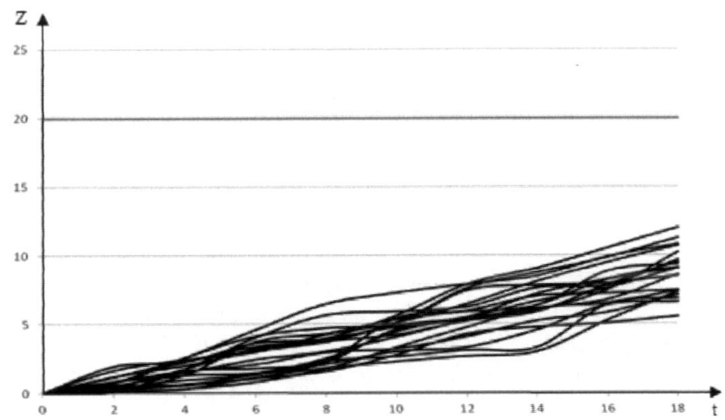

Figure 3. Degradation paths and critical level $z_0 = 20$.

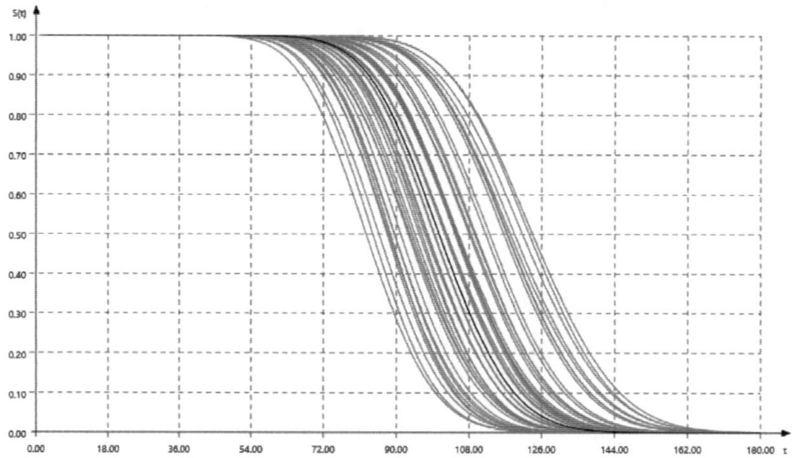

Figure 4. Reliability functions for gamma degradation model with the linear trend and $n = 20$, $k = 10$, $z_0 = 20$.

5. Summary

In this paper, we have analyzed the possibility of reliability estimation in nondestructive testing on the basis of degradation processes. Two widely used degradation models: the gamma and Wiener models have been considered. Summarizing the investigation results, we can conclude that the sample size has a greater influence on the accuracy of the reliability estimates in comparison with the frequency of measuring degradation. Moreover, it has been shown that the more complete degradation paths were observed (the closer degradation index to the critical level), the more accurate reliability estimates will be obtained.

Acknowledgments

The research is supported by the Russian Ministry of Education and Science (project № 1.1009.2017/PCh).

References

[1] Bagdonavičius V *and* Nikulin M 2002 *Accelerated life models: modelling and statistical analysis* (Florida: Chapman & Hall/CRC)
[2] Chimitova E *and* Chetvertakova E 2014 TSU *J of Control and Computer Science* **4(29)** 51–60 Retrieved from http:// journals.tsu.ru/informatics/en/ &journal_page=archive&id=1119&article_id =17728
[3] Chimitova E *and* Chetvertakova E 2015 *Proc. of AMSA* 161 Retrieved from http:// http://www.amsa.conf.nstu.ru/amsa2015/about-workshop/AMSA2015-proceedings.pdf
[4] Liao C M *and* Tseng S – T 2006 *IEEE Trans Reliab* **55(1)** 8840280 DOI: 10.1109/TR.2005.863811
[5] Meeker W Q *and* Escobar LA 1998 *Statistical Methods for Reliability Data* (New York: John Wiley and Sons)
[6] Natalinova N M *et al* 2016 *Proc. of SIBCON* 16090594 DOI: 10.1109/SIBCON.2016.7491866
[7] Natalinova N *et al* 2016 *IOP Conf. Ser.: Mater. Sci. Eng.* **132(1)** 012029 DOI: 10.1088/1757-899X/132/1/012029
[8] Tsai C - C *et al* 2011 *J Stat Plan Inference* **141(12)** 25–35 DOI: 10.1016/j.jspi.2011.06.008
[9] Tsai C - C *et al* 2012 *IEEE Transactions on Reliability* **61(2)** 6198318 DOI: 10.1109/TR.2012.2194351
[10] Whitmore G A 1995 *Lifetime Data Anal* **1(3)** 307–319 DOI: 10.1007/BF00985762

Transformer Oil Dielectric Strength in the Contact Gap of the Explosive Arc-Extinguishing Device

I O Muravlev[1], M A Surkov[1], E V Tarasov[1], Kh Tsoggerel[2], and N F Uvarov[3, 4]

[1]Associate Professor, National Research Tomsk Polytechnic University, Tomsk, Russian Federation
[2]Student, National Research Tomsk Polytechnic University, Tomsk, Russian Federation
[3]Professor, Institute of Solid State Chemistry and Mechanochemistry of the Siberian Branch of the RAS, Novosibirsk, Russian Federation
[4]Professor, Novosibirsk State Technical University, Novosibirsk, Russian Federation

E-mail: tarasovev54@yandex.ru

Abstract: The article describes the experimental results on the breakdown of the high-speed flow of transformer oil. In real conditions, the flow moves in the contact gap of a high-voltage explosive switch with speeds from 67 to 152 m / s. The geometry of the contact gap is sharply inhomogeneous and forms turbulence in the flow zone. In the arc chute medium the air inclusions pass from the dissolved state to the gaseous and the emerging bubbles enter to the electric field. Breakdown occurs, mainly through gas inclusions. In the moment, the gradient of the breakdown voltage is reduced by 91.6% compared to the static state of the oil. The experiments were carried out on the model of a high-voltage explosive switch, connected to the power circuit of the surge generator. The probing of the gap was made by a standard pulse of 1.5 / 50 μs. As a result, the dependences of the gradient of the breakdown voltage on the flow rate of the transformer oil for the usual geometry of the high-voltage explosive switch contact system are constructed.

1. Introduction

The study of arc extinguishing process in the explosive switch model allowed to explain its high current-limiting effect of the high speed arc extinguishing fluid on the electric arc [1]. The flow quickly pulls the arc immediately after the breaking the circuit, and the formation rates of contact gap as well as its length do not affect the growth of arc voltage. In practice this means that the current-limiting capacity remains constant with the increasing of the device rated current, i.e. with the increase in the section and current conductor mass to be destroyed as well as the decrease in the gap formation rate. Explosive switch in its initial condition (Figure 1), where (a) is initial switching stage; (b) 1, 6 are current conducting electrodes; 2 is explosive charge; 3 is current conducted to be destroyed; 4 is pocket air; 5 is casing; 7 is transformer oil flows; 8 is electric series arcs.

The dynamics of the explosive switch contact gap formation is as follows (Figure 1b). When the explosive switch control system receives the signal to turn off and blow up explosive charge the expanding explosion products dramatically increase the pressure in the cavity of the destroying current conductor and the latter, increasing in diameter, moves towards the arc-extinguishing chamber walls. The transformer oil starts filling the gaps between the destroyable current conductor and fixed electrodes. The oil discharge rate varies in the range 70-150 m/s. The arc length is increasing with the approximate rate of $4 \cdot V_n$ for one module of the current conductor being destroyed. This fact explains one of the reasons of current-limitation for the setting level of the device actuation. Another reason is the intense heat removal from the arc column both at the stage of the flow action and due to the high pressure formed when the contact gap reaches its maximum value. The final stage of the high-voltage circuit shutdown process is determined by the recovery speed of contact gap electric strength.

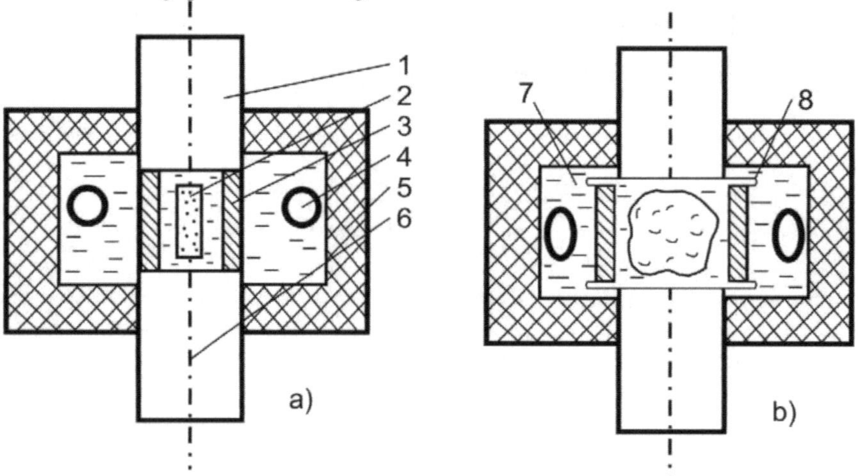

Figure 1. Explosive switch in its initial condition.

There is a way to improve the breaking ability and current limiting capacity of the electric device when a fast-acting fuse with a longer period is connected in parallel to the main current conducting terminals. When the explosive switch is used as a power module, the conductor section can be chosen to be so small that it will be destroyed by the electrical explosion. However in this case it will be necessary to coordinate the voltage surge of the conductor electric explosion with the electric strength of power module period. In the case where a current is switched off or commuted to a parallel circuit for than 70÷120 microseconds, a restrike of the gap may occur in the residual arc channel. But if the time is less than 50÷70 ms then the weakest point is the shortest distance between the current conductor to be destroyed and fixed electrodes where the oil flows. The latter is just the case when a fault current is switched from the fast-acting power module to the parallel-connected high-voltage fuse.

It is known that any fluid perturbation results in a change of its electro-physical properties, e.g. the breakdown voltage [2, 3]. However, quantitative data on the flow dielectric strength of the desired rate range is not found in the literature. This determined the relevance of targeted investigations.

At the same time, occurring vibrations do not have a major impact on the stability of the insulating properties [4].

2. Methods of the Experiment

Experiments on the dielectric strength of the transformer oil flow were performed on a model of the explosive switch contact system, simulating the beginning of a current conductor movement (Figure 2). The installation for the process study allows to register electronic and optical phenomena and consists of a pulsed voltage generator, high-speed photographic apparatus, synchronization and high-voltage measurements devices. The flow probing in the gap was performed by standard positive lightning impulse 1.2/50 ms with a breakdown in the line part of the front if $dU/\ dt = 58\text{kV/ms}$. The breakdown voltage was measured by an active voltage divider and recorded by an oscilloscope. Calibration of the divider and a test of its amplitude and phase distortion are conducted in accordance with the requirements to the measurements of fast processes at high voltage. In the experiments we registered the time span from the beginning of the phenomenon till the breakdown of the studied gap, the breakdown voltage amplitude as well as the optical picture of the process. The synchronization of the elements was provided by the delayed-pulse generator.

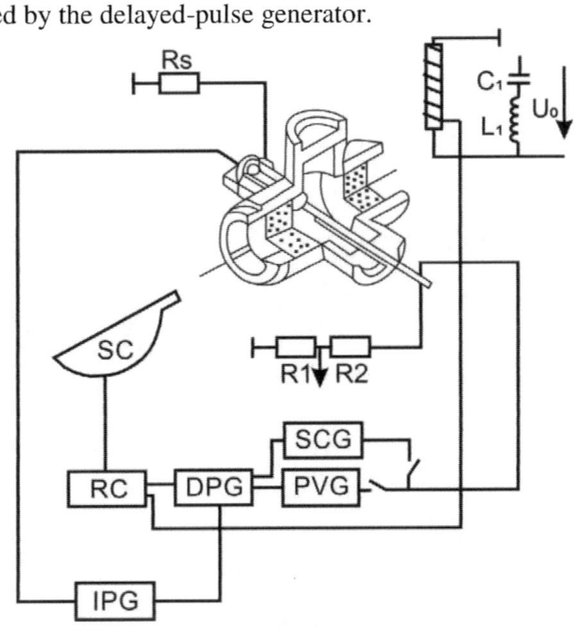

Figure 2. Flow chart of the installation, where SCG is surge-current generator; PVG is pulsed-voltage generator; IPG is initiating pulses generator; DPG is delayed-pulse generator; SC is streak camera; R_1, R_2 are a voltage divider; R_s is shunt; C_2, L_2 are backlight supply; RC is remote control.

The staring point was the explosive charge detonation initiating. The phenomenon was photographed by a high-speed camera (SC) in a slow-motion mode (time lapse) with a frequency of 62.500 frames per second. Recording was conducted in a transmitting light flux. The light was supplied by the impulse discharge lamp powered by a capacitor bank, which in conjunction with the delayed control signal provided the maximum illumination within the studied time span. The phenomenon was recorded on film. To compare the optical detection of flow dynamics with the electrical characteristics we installed the timer – the discharge gap broken down at the time of detonation initiating.

Explosive switch contact gap design is modeled with optically transparent windows that allow to capture images in transmitting light flux. It is presented in Figure 3. The model consists of an explosive 1 and 2 arc-extinguishing chambers filled with transformer oil and simulates the formation of a contact gap in a real device. As in the real apparatus, the model's arc-extinguishing chamber contains air inclusion 5. Explosive charge 3 is located at the end face of the explosion chamber. The model works as follows.

When the charge detonates, expanding explosion products displace oil from the explosion chamber in the arc-extinguishing one. The flow goes to the nozzle opening and wraps around the downstream electrode 4, taking the shape of "umbrella". Hereinafter we call the electrode 4 a lower one. In each experiment, it was fixed and the value of the probed gap remained constant – 1 mm. The measurement of the gap was carried out by the microscope with 2.5% accuracy. The basic experiments used with transformer oil dielectric strength of 20 kV in the standard conditions. Average statistical breakdown voltage excitation pulse was 67 kV to the explosion at an interval of 1 mm.

To control the flow rate in the explosion chamber between the explosive charge and the liquid the metal piston 9 (Figure 3) weighing 50-260 g is installed. With the increase in spraying mass the time of flow acceleration increases whereas its maximum speed reduces correspondingly.

Dependence of breakdown voltage gap on the oil flow rate is shown in Figure 4. It shows that with the velocity Vn increasing in the range of 67 to 152 m / s, U_{bd} is reduced by (70.5–91.6)% compared with the stationary oil state and lies in the range from 19.8 to 5.6 kW . The graph shows the range of the mean root square deviation of the experimental data reaching the maximum value 17% at the point Vn = 75 m / s.

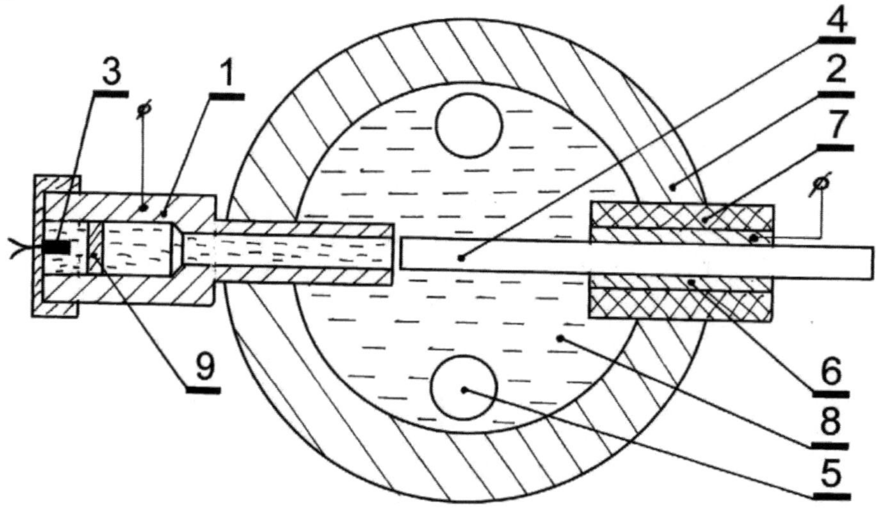

Figure 3. Contact system of the high-voltage explosive switch model, where 1 is explosion chamber; 2 is arc-extinguishing chamber; 3 is explosive charge; 4 is movable rod; 5 is pocket air; 6 is guide, 7 is insulator; 8 is transformer oil; 9 is piston.

Function $U_{np} = f(V_n)$ is close to power dependence and its approximating expression found by the least square method is as follows:

$$U_{bd} = 10^3(223 \cdot V_n^{-0,21} - 72.1) \qquad (1)$$

The reduction of the transformer oil dielectric strength in the flow under the influence of power frequency AC voltage is stated in [2, 3]. According to the authors with speed increasing up to 2 m/s U_{bd} reduction was in 15-35% range depending on the geometry of the electrode system, to a greater or lesser extent increasing the turbulence in the breakdown zone. Reduced strength is explained by the fact that gas or impurities could appear in the gap as well as the influence of dissolved moisture. According to the authors, when oil flows the moisture becomes emulsified and polarized under high voltage contributing to the formation of a weak link.

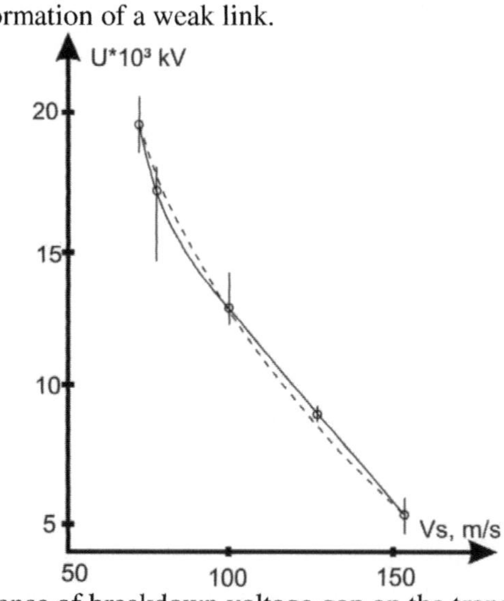

Figure 4. Dependence of breakdown voltage gap on the transformer oil flow rate.

Although conditions [1, 2] differ significantly from the studied in this article, the reduced strength could also be explained by physical processes occurring in the contact gap. Rough explosion chamber walls and friction both in a boundary layer and between layers of fluid give rise to vorticities. Due to the sharp expansion of the flow at the entrance to the arc-extinguishing chamber and flow around the lower electrode the oil in the contact gap becomes intensively turbulent that is clearly visible in the SC-gram. The degree of turbulence is estimated by the dimensionless Reynolds number, which is proportional to the flow rate when the other parameters are constant [5]:

$$\mathrm{Re} = \frac{V_n * \delta}{\lambda}, \tag{2}$$

where V_n is the flow rate; δ is characteristic dimension, in this case, the value of a contact gap; λ is the kinematic viscosity of the medium. In the experiments $3*10^3 \le \mathrm{Re} \le 15*10^3$ and according to [5] the flow relates to the category of high-turbulent ones. It is characterized by chaotic changes of instantaneous speed, pressure and density at various points in the volume. The profile of the explosion chamber outlet and the sharp edges of the lower electrode contribute to the formation and breakdown of vortex on the flow border at the highest Reynolds numbers.

As it is well known [6, 7], transformer oil has complex chemical and fractional composition with a sufficiently high content of dissolved moisture and gases. According to some data the air solubility amounts to 10% of the volume at a temperature of 20 °C and atmospheric pressure. Sudden pressure fluctuations in local flow points with the turbulence in the latter lead to the formation of cavitation bubbles of steam and gas. Thus bubbles in the breakdown region, and in particular, intense cavitation flow along the boundary of the lower electrode, promotes the formation of a weak link in the contact gap. Based on the Reynolds number characteristics bubble inclusions should increase with the degree of turbulence in the flow, and hence the velocity of its movement, which explains the nature of the U_{bd} decline in in the explosive switch gap.

The experimental evaluation of the moisture effect on the dielectric strength of the flow in the breakdown of the "dry" transformer oil, the strength of which under standard conditions was 50 kV, and the average impulse breakdown voltage in the 1 mm gap was 91.5 kV (static). The dielectric strength of the flow, moving with a speed of 152 m/s turned out to be higher by 34%.

Establishing a quantitative relation between the discharge rate of the transformer oil, the value of the contact gap and dielectric strength will allow to define protective characteristic for the explosive switchboard with currentless circuit opening for two-stage explosive switch design. However in the real switch the contact gap increases with time and the value of the breakdown voltage for the current conductor movement may not follow the linear law of distance increasing. To find out how the dynamics of the contact gap formation affects the gradient of the breakdown voltage at the curve points 67 m/s and 152 m/s in Figure 4 a series of experiments for 2 mm gap was conducted. The results showed that within the ranges of actual oil discharge speed and the size of the explosive switch gap on flow action stage the breakdown voltage gradient remains within the limits of the experimental data. Then using an empirical relationship [1] to calculate the oil flow rate in the actual design of the explosive switch, we can state:

$$E_{np} = \{223[0.29D(\frac{m_z}{m_M}) + 0.00289]^{-0.21} - 72.1\} \tag{3}$$

where D is detonation velocity; m_z is explosive charge mass; m_M is sprayed mass, determined by the dimensions of the current conductor.

The results of the investigation were implemented in education process of Tomsk Polytechnic University also [8, 9].

3. Summary

1. The empirical relationship of the impulse breakdown strength for a high-speed flow of transformer oil on the speed of the latter was obtained as a result of investigations.
2. In conjunction with the movement dynamics of the current conductor to be destroyed the relationship allows to calculate the protective characteristics of the explosive switch with currentless circuit opening in the two-stage model of the explosive switch.

References

[1] Korol'kov V. L. *and* Sivkov A. A. 1983 *Elektrotekhnika* **4** 21–23 (in Russian)
[2] Stepanchuk K. F. *and* Tinyakov N. A. 1964 *Energetika* **12** 26–28 (in Russian)
[3] Lashkov S F 1965 *The study of the oil flow dielectric strength. Proceedings of VEI "High Voltage Switchgear"* **72** (Moscow: Energiya) (in Russian)
[4] Mustafina R. M. *et al* 2016 *J. of Physics: Conf. Ser.* **671(1)** 012046. DOI: 10.1088/1742-6596/671/1/012046
[5] Reynolds A J 1979 *Turbulent flows in Engineering* (Moscow: Energiya) (in Russian)
[6] Lipstein R. A. *and* Shakhnovich M. I. 1983 *Transformer oil* (Moscow: Energoatomisdat) (in Russian)
[7] Shakhnovich M. I. 1972 *Synthetic fluid for electrical apparatus* (Moscow: Energiya) (in Russian)
[8] Tarasova E S 2015 *Mediterranean Journal of Social Sciences* **6(3)** 350–355
[9] Daminov I *et al* 2016 *EPJ Web of Conferences* **110** 01015 DOI: 10.1051/epjconf/201611001015

Non-Contact Technique for Determining the Mechanical Stress in thin Films on Wafers by Profiler

N A Djuzhev[1], A A Dedkova[2], E E Gusev[2], M A Makhiboroda[3], and P Y Glagolev[2]

[1]Director R&D center «MST and BEC», National Research University of Electronic Technology «MIET», Moscow, Russia
[2]Engineer NMST, National Research University of Electronic Technology «MIET», Moscow, Russia
[3]Director, NMST, National Research University of Electronic Technology «MIET», Moscow, Russia

E-mail: bubbledouble@mail.ru

Abstract. This paper presents an algorithm for analysis of relief for the purpose of calculating mechanical stresses in a selected direction on the plate in the form of software package Matlab. The method allows for the measurement sample in the local area. It provides a visual representation of the data and allows to get stress distribution on wafer surface. Automated analysis process reduces the likelihood of errors researcher. Achieved time saving during processing results. In carrying out several measurements possible drawing card plate to predict yield crystals. According to this technique done in measurement of mechanical stresses of thermal silicon oxide film on a silicon substrate. Analysis of the results showed objectivity and reliability calculations. This method can be used for selecting the optimal parameters of the material deposition conditions. In software of device–technological simulation TCAD defined process time, temperature and oxidation of the operation of the sample environment for receiving the set value of the dielectric film thickness. Calculated thermal stresses are in the system silicon–silicon oxide. There is a good correlation between numerical simulations and analytical calculation. It is shown that the nature of occurrence of mechanical stress is not limited to the difference of thermal expansion coefficients of materials.

1. Introduction

Technologies of micromechanics or microelectromechanical systems (MEMS) and integrated circuits (IC) develop quickly [1]. The mechanical stresses arising during manufacturing of IC and MEMS devices on Si plates have a strong impact on their reliability and dynamic characteristics [2–4].

Deformation occurs at temperature gradient in the structure, due to the difference between the coefficients of thermal linear expansion of different layers. Magnetostrictive effect, inverse piezoelectric effect or result of action of external forces are reason for strain. As a result, tensile stress in film structure typically results in film cracking or delamination. On the other hand, stress can increase the limit of the system of elasticity, fatigue strength, corrosion–mechanical stability, transistor performance. Therefore, it is important to evaluate and control the amount of stress, not only to avoid damage to device [2, 5], but also to improve product specifications.

The existing techniques for determination of mechanical stresses are based on wafer deformation (use of Newton rings, Stoney formula [2, 6]) or connected with measurement of lattice parameters (Raman spectroscopy [7], X–ray diffractometry [8, 9], transmission electron microscopy).

2. Problem statement

It is necessary to control the amount of stress to optimize the film deposition conditions. The authors calculated mechanical stresses on bending plate. The process of analyzing the surface of a relief on profilometer takes time. This is due to the fact that, to obtain a radius of curvature at the ends of two markers investigated interval must be put. If you move the marker using computer mouse errors may occur. Therefore, there is a need to develop a technique which includes a program with an algorithm relief analysis.

3. Research questions

The study was solved a number of problems. Calculate radius of curvature of the surface based on the model of the geometric representation. Visualize the data obtained. Select a programming language to design a program that implements the algorithm developed. Prove the objectivity and reliability of the presented techniques. Carry out numerical simulations to determine the parameters of the technological process of obtaining the required thickness of the film. Make an experiment. Compare the results of numerical modeling of thermal stress with the analytical calculation. Show different nature of occurrence of stress.

The main objective is to determine amount of stress in the local region of sample, reduction of relief analysis time and minimizing the probability of error.

4. Research methods

In this paper we used Stoney method [2, 6, 10]. Mechanical stresses are calculated by Equation (1) according to the bending wafer (the amount of change of the radius of curvature localized part of a surface of wafer):

$$\sigma f = \frac{E_S \cdot d_S^2}{6(1-u_s)} \cdot \frac{1}{d_f} \cdot \left(\frac{1}{Rafter} - \frac{1}{Rbefore}\right), \quad (1)$$

where σ_f are mechanical stresses; E_s is Young modulus of the substrate material; d_s is thickness of wafer; u_s is Poisson's ratio of the substrate material; d_f is thickness of film on wafer; R_{after} is curvature radius of surface after operation; R_{before} is curvature radius of surface before operation.

Scanning the surface of an optical profilometer is one way of determining the radius of curvature.The result is determined by relief of surface.

In the software for the modern devices are often incorporated in the calculation of the simplification of the stress. That is, the radius of curvature is much greater than the distance between adjacent points [3]. Therefore, by increasing the length of the scanning step error probability occurs. Also, the software of modern optical profilers usually does not provide the ability to automatically calculate the radius of curvature at all points of the surface profile. It became necessary to develop specialized techniques.

The technique involves performing measurements with the non–contact optical profiler and interpretation of the data obtained from it: calculation of the radius of curvature of the surface based on the model representation, and the subsequent calculation of the values of stress at selected area. It achieved a visual representation of the data. First, we analyzed geometry of surface operating plate. Scheme analysis of surface relief is shown in Figure 1.

We conducted geometric transformations, obtained an Equation (2) for calculating radius of curvature of surface of the local area:

$$R=\frac{\left\{\dfrac{(x2-x1)\cdot(y3-y1)}{(x3-x1)\cdot\sin(arctg\dfrac{(x3-x1)}{(y3-y1)})}\right\}^2+\left\{\left\{(y2-y1)-\dfrac{(x2-x1)\cdot(y3-y1)}{(x3-x1)}\right\}\cdot\sin\left\{arctg\dfrac{(x3-x1)}{(y3-y1)}\cdot\right\}\right\}^2}{2\cdot\left\{(y2-y1)-\dfrac{(x2-x1)\cdot(y3-y1)}{(x3-x1)}\right\}\cdot\sin\left\{arctg\dfrac{(x3-x1)}{(y3-y1)}\right\}} \tag{2}$$

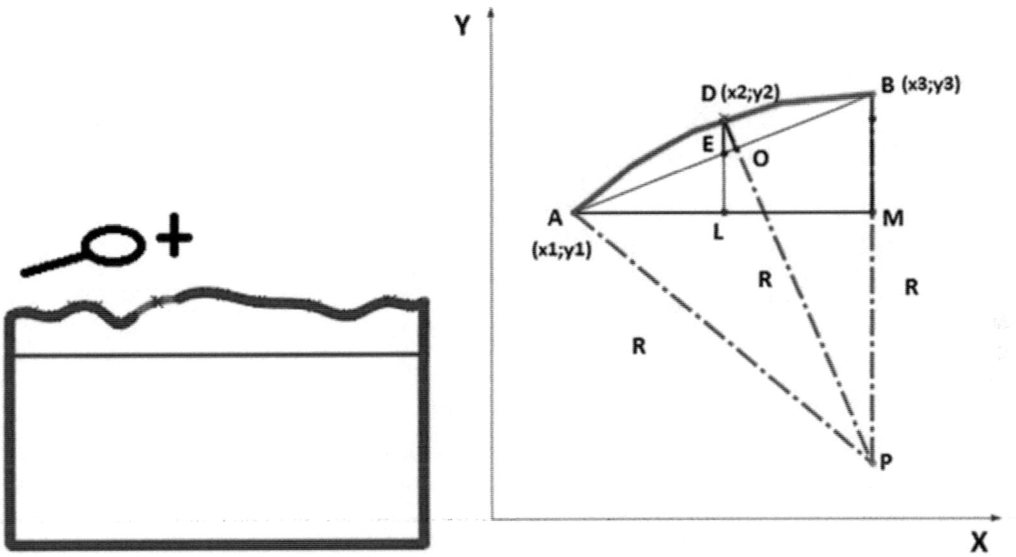

Figure 1. The geometry of surface wafer to calculate radius of curvature.

The reliability of formula (2) tested in SolidWorks software environment.

Built ADB arc with arbitrary values of the coordinates of points (xi,yi). Further, the radius R is made calculation by formula (2). After that, done segments BP and AP with a length R. So that the length of the segment AP, DP, and BP of the same. The closed contour ADBP obtained proves the accuracy of the formula (2).

The second step was to make the algorithm calculating value of the local surface curvature radius by the formula (2) in the Matlab software.

The algorithm consists of several stages. In the first stage polynomial function was found that most correctly describes the relief. Next, we compared the value of the two nearest points of relief. If there is an increase function describing the relief of surface, calculation is made by Equation (2). If the function decreases, the sign of calculated values is reversed. At the end, values of radii are substituted in formula (1) to calculate the mechanical stresses.

5. Findings

The authors investigated a silicon wafer (diameter of 150 mm, thickness of 670 micrometers, orientation (100)) with thermal SiO_2 dielectric film. As is known, SiO_2 membrane is part of air flow sensors in [1]. Modeling was carried out in software of device–technological TCAD simulation. We determined process time, temperature and the operation of the sample oxidation environment for the necessary values of the dielectric film thickness.

The result of the simulation is shown in Figure 2.

In this way, at a temperature of 1000°C the active stage of the process of wet oxidation will be 220 minutes. Thickness of film on experimental sample was determined by Horiba ellipsometer and was 580 nm, which shows a good correlation between the results in TCAD environment and experiment.

Next, using plasma chemical etching without using a mask, silicon oxide film has been removed from back side wafer. A relief measurement in two directions: parallel and perpendicular to base shear before and after deposition. Measurement of relief of a surface was carried out using a non–contact optical profilometer Veeco Wyko NT9300. The results of measuring relief of a surface in perpendicular direction of base shear are shown in Figure 3. We analyzed the work area from 20 to 130 millimeters.

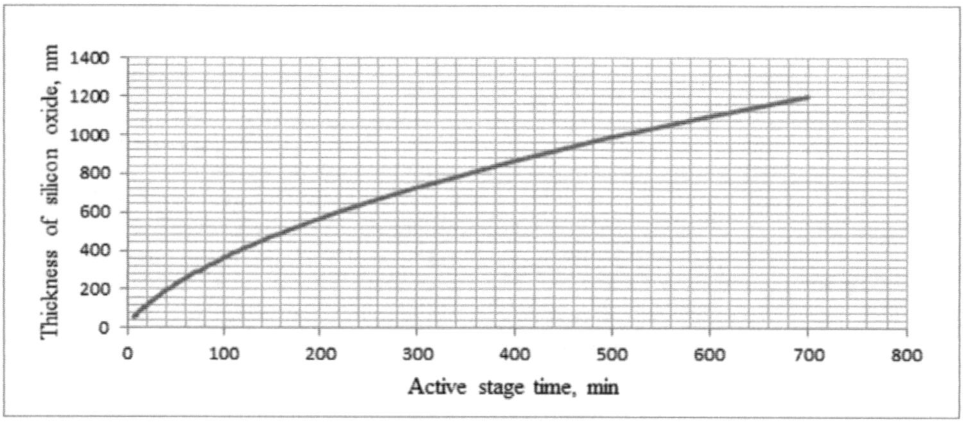

Figure 2. The dependence of film thickness of active stage of process time.

From these figure it can be concluded that after forming silicon oxide film on front side of plate became more curved. As is known, initial wafer of monocrystalline silicon with orientation (100) contains minor positive mechanical stresses. Hence, value of stress increased. For the equilibrium state of the substrate-film, resulting force compression-tension must be equal to zero. Therefore, thermal SiO_2 film negative compressive stresses are equal in magnitude stresses in silicon.

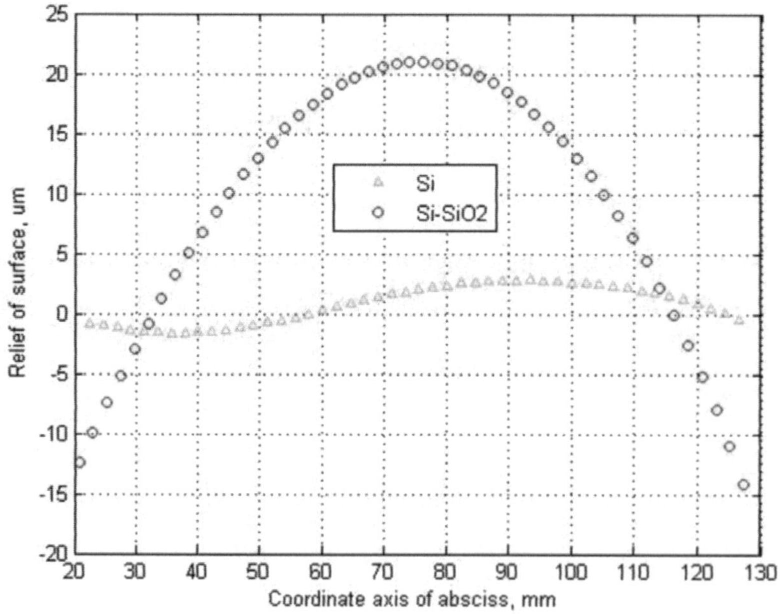

Figure 3. Surface profile in a perpendicular base shear direction.

With above-described algorithm in Matlab calculation performed curvature values before and after film deposition. Further, by formula (1) using the algorithm in Matlab determined magnitude of stress in the structure (Figure 4) after forming the SiO_2 film of 600 nm on the front side. Constant biaxial modulus ($E_s/(1\text{-}u_s)$) of the sample in the crystallographic plane (100) was $1.8\cdot10^{11}$ Pa [10, 11].

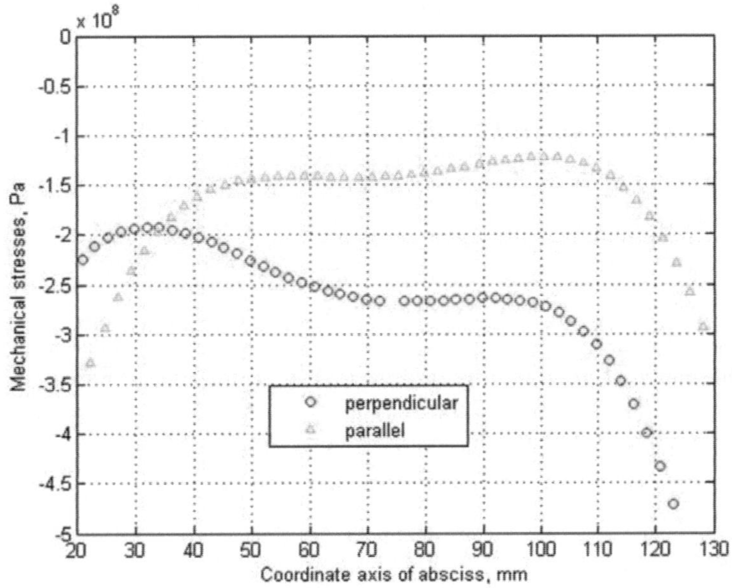

Figure 4. The distribution of mechanical stresses in the wafer

Figure 4 shows that at edges wafer increases the stresses. The average value mechanical stresses on the work area substrate is - 210 MPa, which corresponds to the order of magnitude [2, 3, 7, 10]. Comparing the values obtained by the maximum stress and the critical stress for the material, it is possible to prevent cracking of the structure. For example, by making structure of heat treatment the purpose of minimization stress [7, 12]. To determine nature of occurrence of mechanical stresses carried out comparative analysis of stress values and value of thermal stress calculated in TCAD software.

6. Numerical simulation of thermal stress in TCAD.

Presenting results of simulation of thermal stress in the silicon oxide film on a silicon substrate with crystallographic orientation (100) on the basis of device–technological TCAD simulation software package.

In TCAD software section process does not provide for setting value of Young's modulus E_f and Poisson's ratio u_f film. However, assigning a value of module substrate K_f and the shear modulus G_f of film material, it is possible to calculate the required values, determined system of Equations (3):

$$\begin{cases} K_f = \dfrac{E_f}{3(1-2\cdot u_f)} \\ G_f = \dfrac{E_f}{2(1+u_f)} \end{cases} \quad (3)$$

Material parameters used in TCAD library: value of substrate module K_f is $6.534\cdot10^{10}$ (Pa) and shear modulus G_f $2.792\cdot10^{10}$ (Pa), respectively. Solving the system of Equations (3), received the Young's modulus E_f of SiO_2, equal to $8\cdot10^{10}$ (Pa). In the process of calculating difference in TCAD temperature coefficient of linear expansion of material $\Delta\alpha$ silicon and silicon oxide with a crystallographic orientation (100) was assumed to be - $1.5\cdot10^{-6}$ (1/°C) [3]. The magnitude of difference of film deposition temperature SiO_2 and measurement is ΔT 1000°C. As a result of simulation obtained stress value in the silicon oxide film -110 (MPa).

The purpose of analyzing correlation of numerical simulation results with analytical calculation, calculated the value σ_{term} thermal stress. In the analysis of one–dimensional problem with an absence of cross–compression Poisson ratio u_f tend to zero. Obtained Equation (4) to simplify the expression of the work [10]:

$$\sigma_{term} = E_f \cdot \Delta\alpha \cdot \Delta T \quad (4)$$

The calculated value of thermal stress in SiO_2 film was - 120 MPa.

Analyzing the results of numerical simulations and analytical calculations, we can conclude about good comparability of results. Thus obtained value of thermal stress is less than the mechanical stresses obtained experimentally using techniques described. The difference with experimental data indicates the multifaceted nature of stress, which includes the thermal stress, but not limited thereto.

7. Conclusion

Presented and implemented a technique that allows for analysis of surface relief for purpose of calculation stress. The method allows to obtain distribution of mechanical stresses on wafer surface. Analysis of the results showed objectivity and reliability calculations. This method can be used for selecting the optimal parameters of deposition material conditions.

8. Acknowledgement

This work was supported by the Ministry of Education and Science of the Russian Federation (within the agreement № 14.578.21.0188, id RFMEF57816X0188).

References

[1] Djuzhev N A *et al* 2015 *Proc. of Universities. Electronics* **20(6)** 644 (in Russian)
[2] Lindroos V *et al* 2010 *Handbook of Silicon Based MEMS Materials and Technologies*, (Burlington: Elsevier) ISBN 978-0-8155-1594-4
[3] Laconte J *et al* 2006 *Micromachined Thin-Film Sensors for SOI-CMOS Co-Integration* (US: Springer) DOI: 10.1007/0-387-28843-0
[4] Leondes C T 2006 *Mems/Nems Handbook* (US: Springer) DOI: 10.1007/b136111
[5] Petersen K E 1982 *Proc. IEEE* **70(5)** 420 DOI: 10.1109/PROC.1982.12331
[6] Stoney G G 1909 *Proc. Royal Society A* **82(553)** 172 DOI: 10.1098/rspa.1909.0021
[7] Fitch J T *et al* 1989 *J. Vac. Sci. Technol., B* **7(4)** 775 DOI: 10.1116/1.584599
[8] Stanojevic A *et al* 2016 *IOP Conf. Ser.: Mater. Sci. Eng.* **119** 012007 DOI: 10.1088/1757-899X/119/1/012007
[9] Vourna P 2016 *IOP Conf. Ser.: Mater. Sci. Eng.* **108** 012017 DOI:10.1088/1757-899X/108/1/012017
[10] Kobeda E *and* Irene E A 1985 *J. Vac. Sci. Technol., B* **4(3)** DOI: 10.1116/1.583603
[11] Sinha A K *et al 1978* *J. Appl. Phys.* **49(4)** DOI: 10.1063/1.325084
[12] Guan D *et al* 2014 *JMM* **24(2)** 027001 DOI:10.1088/0960-1317/24/2/027001

The application of the Methods of Cathodoluminescence and Photoluminescence for Non-destructive Testing of AlGaN Heterostructures

Y A Kontsevoy[1], E Ph Pevtsov[2], R A Khmelnitskiy[3], Y V Kolkovskiy[1], V V Gruzdov[4], and S A Malykhin[5]

[1]Principal Researcher of "PULSAR" S&PE" JSC, Moscow, Russia
[2]Director of Design Center of Integrated Circuits, Nanoelectronics Devices and Microsystems, Moscow Technological University (MIREA), Russia
[3]Leading Researcher of P.N. Lebedev Physical Institute of the Russian Academy of Sciences Moscow, Russia
[4]General Director of JSC "PULSAR"S&PE", Moscow, Russia
[5]Post-graduate, Faculty of Physics M.V. Lomonosov Moscow State University, Russia

E-mail: pevtsov@mirea.ru

Abstract. This paper examines cathodoluminescence spectra of samples on sapphire substrates to develop methods of non-destructive testing of wafers with AlGaN/GaN heterostructures. It has been determined that the cathodoluminescence peak of AlGaN compound was demonstrated for decreased energy of excitation electrons (at 0.5 eV and 1 keV) only. Cathodoluminescence peak of AlN compound with energy of 6.15 eV was demonstrated at any excitation energy of 0.5 to 6.15 eV. It has been demonstrated that cathodoluminescence spectrum analysis allowed determining aluminum percentage in AlGaN, which was essential for inward testing of wafers for microwave transistors production. Data on the intensity distribution of "yellow" photoluminescence over the wafer surface has been obtained, and it has been demonstrated that these measurements were a prospective method of non-destructive testing to sorting wafers by defects.

1. Introduction

There are a number of publications which state that in order to obtain high-quality AlGaN/GaN-based high electron mobility transistors, Al content in the AlGaN layer should be within the range of 20 to 35 %. The main purpose of this work is to develop cathodoluminescence methods as implements for non-destructive testing in order to determine aluminum content in heterostructures before their production is started. Methods of cathodoluminescence and photoluminescence are effectively used for examination of properties of wide-band gap materials and structures, in particular, for examination of dislocation nature in GaN thin films and for control of self-radiation related to a band-to-band junction [1–3]. Cathodoluminescence and photoluminescence allow detecting the defects related to luminescence in the visible spectrum. The relation between the so-called "yellow" cathodoluminescence excited by electrons of various energy and resistance of ohmic contacts of a GaN source and drain of microwave transistors has been ascertained. Aluminum content in AlGaN layers has been examined, in particular, in the work [3], however, thick layers of specified material (hundreds of nanometers thick) have been used. In this work, AlGaN layers with a thickness of 20nm have been examined.

IOP Conf. Series: Materials Science and Engineering **189** (2017) 012020 doi:10.1088/1757-899X/189/1/012020

2. Examination of Cathodoluminescence

JSM7001F scanning electronic microscope (JEOL, Japan) has been used for the purpose of the examination. Electronic microscopic images have been obtained in secondary electrons for electron energy of 15 keV. INCA Energy 350XT add-on device (Oxford Instruments Analytical, UK) has been used for the purpose of performing X-ray fluorescent analysis. Thus, peak x-ray characteristic radiation of film elements has been analyzed. MonoCL3+ add-on device for spectral analysis of cathodoluminescence (Gatan, USA) has been used for the purpose of the cathodoluminescent examination. The light induced in near surface layers of the sample affected by energetic electrons has been collected with parabolic mirror and directed to the input slot of a monochromator equipped with rotary diffraction grating. Cooled photoelectronic multiplier has been used as a light receiver. Spectral range of the spectrometer is 190 to 1100 nm. Spectra obtained have not been adjusted to the spectrometer's performance. 1200 grooves/mm diffraction grating with maximum shining at a wavelength of 500 nm has been used to record reviewed (within the range of 190 to 800 nm) spectra of GaN/AlGaN structures; 1200 grooves/mm diffraction grating with maximum shining at a wavelength of 250 nm is used to record UV range spectra. The sample has been cooled down to 80 K temperature, but most of the spectra have been recorded at room temperature. Spectral resolution has been 1nm. It has been established that most informative were the cathodoluminescence spectra measured for electron energy of 5, 3, 2, 1 and 0.5 keV. Figure 1 illustrates the graphical representation of the excitation area depth for such energies.

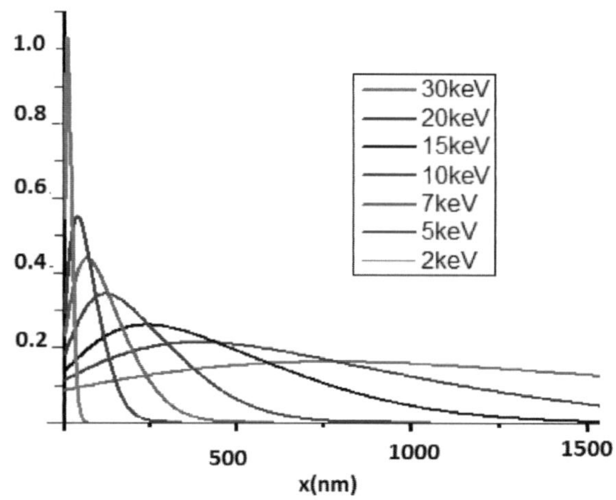

Figure 1. Assessment of the energy loss per length unit depending on the penetration distance of electrons for various energy of electrons (Monte-Carlo simulation).

The structures examined in our work are the following (as from surface): i-GaN (1.5 nm), i-AlGaN (20 nm), i-GaN (2300 nm), nucleation layer (thickness is not specified), sapphire. According to the manufacturer's data, Al content in structure of $Al_xGa_{1-x}N_x = 0.3$.

Figure 2 and Figure 3 illustrate the graphical representation of the cathodoluminescence spectra for various energy of electrons exciting cathodoluminescence. Figure 2A and Figure 3a clearly illustrate the highest peak of cathodoluminescence related to GaN excitons (3.42 keV).

The peak with energy of 6.15 eV is demonstrated for all energies of exciting electrons, which is approximately corresponding to the forbidden band of AlN. Nevertheless, location of the AlN layer is not clear, and the manufacturer has not mentioned this layer. Furthermore, there is no peak related to the AlGaN layer for exciting electron energy of 3 keV. There are no such peaks for higher energy of electrons exciting cathodoluminescence up to 15 keV. In addition, the cathodoluminescence peak with energy of 5.12 eV is indicated for all energies of exciting electrons. Unfortunately, this peak is hard to identify, possibly, it is related to junctions in the forbidden band of the AlN wide-band material.

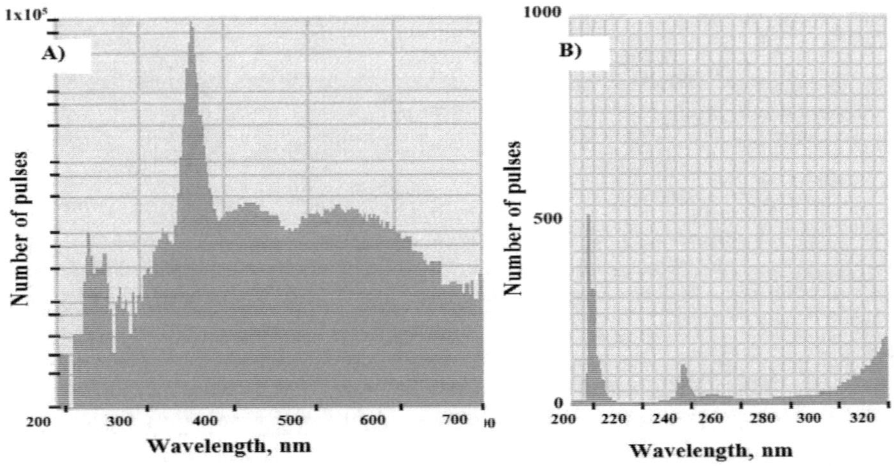

Figure 2. Cathodoluminescence spectra, where A) in logarithmic scale; B) cathodoluminescence peaks in the UV spectrum in a linear scale. Energy of exciting electrons is 3 keV, current is 46 nA.

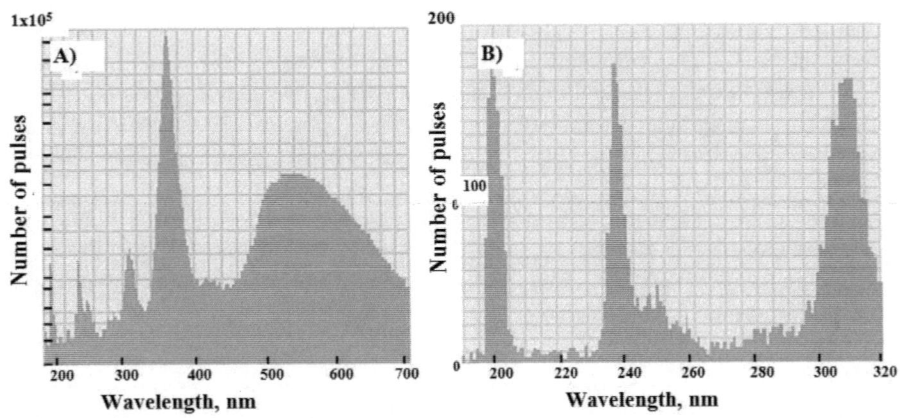

Figure 3. Cathodoluminescence spectra, where A) in logarithmic scale; B) cathodoluminescence peaks in the UV spectrum in a linear scale. Energy of electrons is 0.5 keV, current is 19 nA.

The dependence of the maximum position of the exciton photoluminescence peak for triple compound $Al_xGa_{1-x}N$ on x (atomic fraction of Al) has been determined by experiment in the work [3, 4]. In this case, the cathodoluminescence peak with energy of 3.97 eV has been demonstrated for energy of exciting electrons of 0.5 eV (and for energy of 1 keV) which corresponded to aluminum content of $x = 0.24$. It should be noted that, the maximum value of "yellow" cathodoluminescence for decreased energy or exciting electrons is very clearly indicated in Figure 2A.

3. Examination of Photoluminescence

Photoluminescence, which was excited by ultraviolet LEDs with a wavelength of 295 nm, has been measured on the same samples on which cathodoluminescence has been measured in addition to the examination of cathodoluminescence. The purpose of this work was to obtain a new data on defects in GaN-based structures by photoluminescence methods and develop examination methods and method of inward express-testing of wafers with heterostructures before starting their production. Photoluminescence has been used to make the "maps" of "yellow" photoluminescence over the area of AlGaN/GaN/SiC and AlGaN/GaN/sapphire heterostructures [4–8]. Classic spectrum of "yellow" photoluminescence has been observed, and great variation of a signal over the area of samples related to yellow photoluminescence has been obtained. Photoluminescence level of samples on substrates of silicone carbide were more heterogeneous than those of heterostructure samples on sapphire substrates. The cause of the photoluminescence heterogeneity in samples has not been still identified. However, the photoluminescence heterogeneity is an evidence of heterogeneous properties of heterostructures over their area. The peaks of photoluminescence related to band-to-band junctions have not been detected for GaN and AlGaN layers, which may be explained by insufficient intensity of sample exposure to LED radiation.

Based on the data obtained for inward non-destructive testing of wafers, the device for automatic drawing of "photoluminescence maps" over the sample area [6] has been developed and commissioned. It allows monitoring of 48 sections of photoluminescence for heterostructures with a diameter of 2 inches or 108 sections of photoluminescence for heterostructures with a diameter of 3 inches.

4. Conclusion

When recording cathodoluminescence in AlGaN/GaN heteroepitaxial structures at room temperature, selection of energy of exciting electrons allows clearly identifying spectrum components of exciton radiation even from thin layers of AlGaN and AlN, which allows determining the composition of triple compound of AlGaN.

High heterogeneity of AlGaN/GaN structures has been proven by the fact that the intensity and ratio of exciton radiation components remained almost unchanged over the wafer area.

Effective method of assessment of AlGaN and AlN layer location depth in structures is a variation of energy of a primary electron beam.

Wide bands of 480 to 700 nm corresponding to "yellow" luminescence with a maximum value near 540 nm have been observed in all AlGaN/GaN structures, which was caused by defects in GaN layers.

Radiation band from an optical center with a zero phonon line of 237 nm (quantum energy is 5.2 eV) and the phonon repetition range of 243 to 266 nm has been observed in luminescence spectra. Existence of this band is apparently related to defects of the AlN layer.

"Yellow" luminescence related to defects of wafers with AlGaN/GaN heterostructures was clearly visible in photoluminescence spectra excited by UV LED, which justified the use of a photoluminescence method for quality control of original wafers by recording of "photoluminescence maps."

Acknowledgments

Research was supported by the Ministry of Education and Science of the Russian Federation (No 8.5098).

References

[1] Luchnikov P A 2015 *Russian Physics Journal* **57(9)** 1239–1245 DOI: 10.1007/s11182-015-0369-0
[2] Ragachev A A *et al* 2015 *Appl. Surf. Sci.* **351(1)** 811–818 DOI: 10.1016/j.apsusc.2015.06.008
[3] Luchnikov P A *et al* 2016 *IOP Conf. Series: Materials Science and Engineering* **110(1)** 012081 DOI: 10.1088/1757-899X/110/1/012081
[4] Vasileva O V *et al* 2014 *J. of Physics: Conf. Ser.* **552(1)** 012026 DOI: 10.1088/1742-6596/552/1/012026
[5] Surzhikova O A 2015 *IOP Conf. Ser.: Mater. Sci. Eng.* **81(1)** 012098 DOI: 10.1088/1757-899X/81/1/012098
[6] Yoshida S *et al* 1982 *Journal of Applied Physics* **53(10)** 6844–6848 DOI: 10.1063/1.329998
[7] Pevtsov E Ph *et al* 2017 *IOP Conf. Ser.: Mater. Sci. Eng.* **168(1)** 012095 DOI:10.1088/1757-899X/168/1/012095
[8] Gladysheva N B *et al* 2015 *Materials of Electronics Engineering* **18(2)** 146–148 DOI: 10.17073/1609-3577-2015-2-146-148 (in Russian)

IOP Conf. Series: Materials Science and Engineering **189** (2017) 012021 doi:10.1088/1757-899X/189/1/012021

X-Ray Diffraction Analysis of Bottom Ash Waste after Plasma Treatment

G Volokitin[1], Yu Abzaev[1], N Skripnikova[1], O Volokitin[2], and V Shekhovtsov[3]

[1]Professor, Tomsk State University of Architecture and Building, Tomsk, Russia
[2]Associate Professor, Tomsk State University of Architecture and Building, Tomsk, Russia
[3]Graduate Student, Tomsk State University of Architecture and Building, Tomsk, Russia
[4]Student, Tomsk State University of Architecture and Building, Tomsk, Russia

E-mail: volokitin_oleg@mail.ru

Abstract. The paper deals with the plasma-chemical synthesis of melts produced from the bottom ash waste for the production of new construction materials with enhanced performance characteristics. Phase composition of the plasma-treated bottom ash waste is detected by the X-ray diffraction analysis. The bottom ash waste is a mixture of SiO_2 minerals. The structure and phase composition of this mixture are investigated after the plasma treatment. The obtained results are compared with the original state of the mixture. The identification and the qualitative content of ash waste as a multi-phase system are complicated by the overlapped reflections and a possible existence of the intermediate amorphous phase.

1. Introduction

The treatment level of industrial solid waste in thermal power plants (TPP) is, currently, extremely low that leads to a considerable accumulation of bottom ash waste in ash-disposal areas [1–3]. The core aspect of ash waste utilization is a high melting temperature (up to 1700°C) which depends on heterogeneity of its chemical composition. Knowledge of the structural composition of the amorphous and crystal ash phases is very useful for the provision of the effective performance characteristics. To the authors's knowledge, little publications are available in the literature that discuss investigations of the plasma treatment effect on the ash structure and phase composition.

The aim of this work is to study the processes in plasma-chemical reactor occurred during the production of ash-based silicate melt and the identification of its amorphous state after the plasma treatment.

2. Materials and methods

The bottom ash waste fraction (<150 μm) generated by Tomsk TPP–2, was used as a raw material for the production of the silicate melt. The original chemical composition of this material is presented in Table 1.

Table 1 indicates that the chemical composition of the investigated material contains 51% SiO_2. This content is similar to that of commercial glass and can be utilized to produce silicate melts, including the production of mineral fibers [4–6].

IOP Conf. Series: Materials Science and Engineering **189** (2017) 012021 doi:10.1088/1757-899X/189/1/012021

Table 1. The original chemical composition of bottom ash waste generated by Tomsk TPP–2.

Material	Oxide content, (wt.%)					
	SiO_2	Al_2O_3	Fe_2O_3	CaO	MgO	Ignition loss
Bottom ash waste	51.16	34.57	3.62	8.33	0.91	1.41

The XRD analysis was carried out by DRON-4-07 diffractometer which was modified for digital signal processing. Measurements were conducted using copper radiation (K_α) and Bragg-Brentano X-ray optical scheme. Specifications for the DRON-4-07 included 0.02^0 scanning step and $16.0–92.0^0$ range for angles to be scanned. The XRD analysis of obtained melts was based on the Rietveld refinement method [7-8]. This method is used to detect the relative phase content in compliance with the reference structure; space groups; refine the crystal parameters and the space distribution of atoms in crystal lattices. Phase transformations, chemical reactions occurring in the material under heating and cooling conditions were measured in air on the simultaneous thermal analyzer Netzsch STA 449C Jupiter (Germany) at a heating rate of 10 degrees per minute.

3. Experimental

The bottom ash waste was subjected to melting in a plasma-chemical reactor presented in [9, 10]. Its specifications included U=160 V, I=220 A, P=35.2 kW, $q=1.8\cdot10^6$ W/m^2. The operating principle of the plasma-chemical reactor is based on the interaction between the highly concentrated plasma flows and the silicate-containing powdered material. As a result of this interaction, fine particles were heated with the following formation of a homogeneous melt. The configuration of the plasma-chemical reactor allowed eliminating the loss of fine particles blowing by the plasma flow out of the melting zone. Moreover, the obtained silicate melt was homogeneous within the whole volume of the melting furnace.

4. Results and discussions

4.1. X-ray diffraction analysis

Upon completion of the experiments in plasma treatment of high-temperature silicate melts, the X-ray diffraction (XRD) analysis was performed for ash powders both before and after plasma treatment. To identify the unknown phases in the studied substances, it is advisable to address to the Crystallography Open Database [11]. The modeling of amorphous states allows identifying the quantitative phase content in ash waste after the plasma treatment. The amorphous phase predominates in the integral intensity. The number of amorphous phases is obtained on the basis of crystal lattices of ash waste in its original state. Table 2 summarizes the stable phases presented in the ash waste, namely: SiO_2, Al_2O_3, Fe_2O_3, TiO_2, CaO, MgO.

This phase composition is a complex mixture of SiO_2, Al_2O_3, and Fe_2O_3. The amount of SiO_2 phase is the largest and equals to ~56.2%, while the amount of Al_2O_3 and Fe_2O_3 phases is respectively 14.41 and 17.55%.

The angular dependence of the reflected X-ray intensity is determined for each of the reference phases. The Rietveld refinement is used to detect the background, profile, and structural parameters of the intensity of reference phases. The degree of proximity of the estimated intensity with the experimental is determined by R_{wp} convergence criterion. The Rietveld refinement results in the detection of the space distribution of atoms in crystal lattices and the parameters of supercells of the investigated phases. The Rietveld refinement method was implemented using the software described in [5]. Figure 1 contains the plot of the XRD patterns for the certain phases.

The diffraction pattern obtained for ash waste is complex since it includes both the strong and weak overlapping reflections. The convergence criterion for the ash original state is R_{wp} =5.508%.

The Rietveld method allows the authors to list the reference phases in ash waste which contributes to the integral (experimental) intensity.

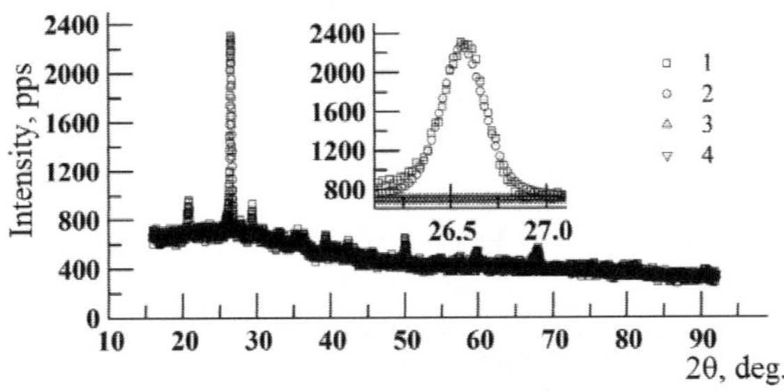

Figure 1. XRD patterns for the original state of ash waste, where experimental (1); SiO_2 (2); Fe_2O_3 (3); TiO_2 (4).

The results of the analysis, the discovered major phases and their contribution to the integral intensity in the oxide mixture are given in Table 2. The XRD analysis shows that superposition of the integral intensity is 94.47%. This amount indicates to the proper identification of phases in ash waste.

Table 2. Structure and phase content refined by the Rietveld method.

N	Chemical formula	Intensity, [%]	*a*, [nm]	*b*, [nm]	*c*, [nm]	α [deg.]	β [deg.]	γ [deg.]
1	SiO_2	56.23	0.49249	0.492491	0.542288	90	90	120
2	Al_2O_3	6.2	0.474452	0.474452	1.283357	90	90	120
3	Al_2O_3	14.41	0.513593	0.513593	0.513593	57.47	57.47	57.47
4	CaO	0.06	0.4799	0.4799	0.4799	90	90	90
5	MgO	0.01	0.4217	0.4217	0.4217	90	90	90
6	Fe_2O_3	0.01	0.50355	0.50355	1.37471	90	90	120
7	Fe_2O_3	17.55	0.50355	0.503553	1.327161	90	90	120
8	TiO_2	<0.01	0.4991	0.4991	0.2879	90	90	120
9	Total	94.47						

As shown in Figure 2, ash becomes amorphous after the plasma-chemical treatment. In order to identify the amorphous phases and their content using the Rietveld method, the amorphous states are simulated for SiO_2, Al_2O_3 and Fe_2O_3 phases which make the main contribution to the intensity in the original state. The calculated atomic density determines the size of the original cubic cell, in which the oxide atoms are concentrated.

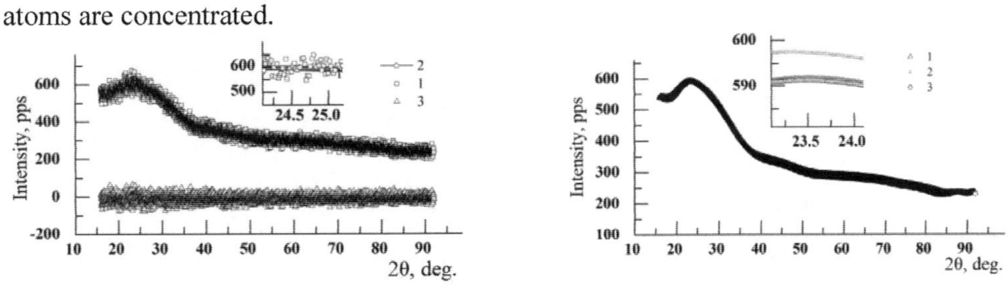

a) *b)*

Figure 2. XRD patterns for plasma-treated ash waste, where *a)* experimental (1); estimated (2); difference (3); *b)* amorphous phases Fe_2O_3 (1), $O_{22}Al_{20}$ (2), SiO_2 (3).

The optimized sizes of the cubic cells and the force-equilibrium location of atoms were then ascertained. As a result, the amorphous state of Fe_2O_3, $O_{22}Al_{20}$ and O_2Si phases was achieved as well as the space distribution of atoms in optimized cubic cells. In case these of the assignment a primary space group to cubic cells, they can be identified as the unit cells of Fe_2O_3, $O_{22}Al_{20}$ and O_2Si. For the optimized cubic cells of $O_{22}Al_{20}$ and O_2Si, the structural factors and consequently integral intensities can be calculated. The model amorphous phases of Fe_2O_3, $O_{22}Al_{20}$ and O_2Si are used to obtain the phase composition of ash waste. Figure 2*b* shows the XRD patterns of plasma-treated ash waste analyzed by the Rietveld method. The convergence criterion is $R_{wp} = 5.31\%$. The refined parameters of cubic cells and the intensity of amorphous phases are presented in Table 3.

Table 3. Amorphous structure and phase content.

N	Chemical formula	Intensity, %	*a*, (nm)	*b*, (nm)	*c*, (nm)	α (deg.)	β (deg.)	γ (deg.)
1	$O_{22}Al_{20}$	43.55	1.3909	1.0361	1.3038	77.08	91.18	81.24
2	$O_{22}Fe_{24}$	49.05	0.9750	1.3209	0.8267	73.69	80.40	100.10
3	$O_{48}Si_{24}$	0.53	1.3364	1.6277	1.2900	78.44	86.10	69.54
4	Total	93.13						

The integral intensity superposition of the amorphous phases is 93.13% of the XRD diffraction pattern plotted in Figure 2*b*. The obtained results indicate that the plasma-chemical treatment leads to a complete amorphization of ash waste. At the same time, the ash melting results in a complex redistribution of the amorphous phase content. Thus, the amount of O_2Si phase decreases, whereas that of $O_{22}Al_{20}$ and Fe_2O_3 decreases. The density of Al and O elements in $O_{22}Al_{20}$ cubic cell changes insignificantly.

4.2. Differential thermal analysis

The differential thermal analysis (DTA) shown in Figure 3, describes the original state of ash waste. The DTA curve indicates that the ash behavior under the heating process is similar to that of classical silicate materials.

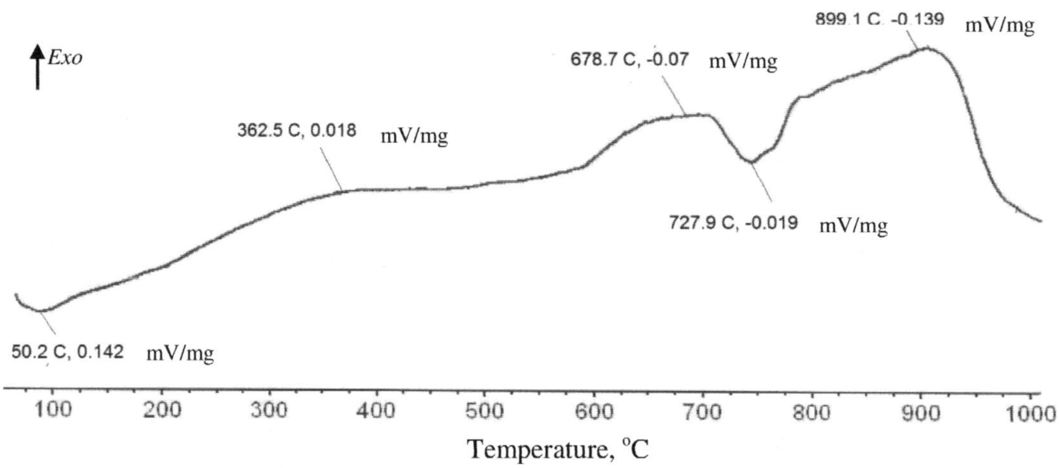

Figure 3. DTA curve for the original state of ash waste.

Within the temperature of 100 °C, the observed endothermic process corresponds to the removal of free water. A blurred exothermic process within 300÷400°C range is caused by the formation of crystallization centers. A further temperature growth up to 600°C leads to the removal of adsorption and chemically bound water. The exothermic process of 600÷700 °C relates to the transformation of the amorphous phase to a crystal. Within 800÷950°C, the exothermic process is caused by softening of low-melting compounds and recrystallization of calcium and magnesian silicates with a complex composition.

5. Conclusions

The physicochemical investigations showed that the low-temperature plasma treatment of ash-based silicate melt resulted in the ordered structure of aluminosolicate glass. The experiments allowed the authors to obtain consistent results on the ash waste phase composition both before and after the plasma treatment. SiO_2 mineral predominated in the ash waste. The plasma-chemical treatment resulted in amorphization of the ash waste which then transformed to a mixture of $O_{22}Al_{20}$, Fe_2O_3 and O_2Si phases. Melting products were homogeneous and ordered, *i.e.* the melting process of the ash waste promoted a complete homogenization of the melt. Therefore, the obtained aluminosolicate glass can be used to produce construction materials, including mineral fibers having the enhanced performance characteristics (chemical, thermal and water resistance), mechanical-and-physical properties and service life.

References

[1] Ángeles G *et al* 2001 *Anal. Chem.* **73(2)** 151 – 156 DOI: 10.1021/ac0006674

[2] Kastner J R *et al* 2005 *Environ. Sci. Technol.* **39(6)** 1835–1842 DOI: 10.1021/es0499492

[3] Gupta N *et al* 2004 *Applied Science and Manufacturing* **35(1)** 103–111 DOI:10.1016/j.compositesa.2003.08.001

[4] Volokitin G G *et al* 2011 *Glass and Ceramics* **68(7-8)** 239–241 (in Russian)

[5] Predoi D *et al* 2016 *Materials* **9(9)** DOI: 778 10.3390/ma9090778

[6] Kim J K *et al* 2000 *Metall and Mat Trans A* **31(4)** 1283–1293 DOI: 10.1007/s11661-000-0123-1

[7] Abzaev Yu A *et al* 2012 *Vestnik of the TGASU* **12(4)** 200–209 (in Russian)

[8] Smolyanskiy V et al 2017 *Web of Conferences* **102** 01033 DOI: 10.1051/matecconf/201710201033

[9] Vlasov V A *et al* 2015 *J. Phys.: Conf. Series* **652** 012031 DOI: 10.1088/1742-6596/652/1/012031

[10] Surzhikov *et al* 2016 *IOP Conf. Ser.: Mater. Sci. Eng.* **110** 012002 DOI: 10.1088/1757-899X/110/1/012002

[11] *Crystallography Open Database* Retrieved from http://www.crystallography.net/cod/

Calculation of Optimal Geometrical Magnification and Spatial Resolution of Betatron Tomograph

Y Zhong[1], S V Chaklov[2], and V B Trinh[1]

[1]Postgraduate Student, Tomsk Polytechnic University, Tomsk, Russia
[2]Head of Russian-Chinese Laboratory of Radiation Control and Inspection, Tomsk Polytechnic University, Tomsk, Russia

E-mail: yan.tchzhun@yandex.ru

Abstract. One of the perspective directions of development of non-destructive testing is the method of computed tomography. Computed tomography really enhances the ability of X-ray inspection, from thin and simple to thick and complex parts. There are many factors that influence the performance of computed tomography, the main parameters for computed tomography scanners and also scanner based on betatron, are geometric magnification and spatial resolution. Calculations of these parameters for the betatron tomograph are shown in this paper.

1. Introduction

Currently, X-ray tubes and linear accelerators are widely used as X-ray sources for industrial CT. Because of the low intensity, betatron is rarely used for industrial computed tomography (CT) [1], although it is widely used as a radiation source in the non-destructive testing [2-3] and successfully used in the inspection system [4]. Only in recent years, the company "PROMINTRO" and the Federal Research and Production Center "Altai" began to develop CT system with the use of betatron, the properties of which was improved and manufactured at the Institute of non-destructive testing of Tomsk Polytechnic University (INDT TPU). Advantages of betatron, if compares with linear accelerators with the same radiation energy, the cost of betatron is 3-4 times lower, and the focal spot size of betatron is much smaller.

Since the refine betatron characteristics were improved, so it become reality and possible to use betatron for industrial CT as a radiation source. Development of industrial CT based on betatron was carried out in "Russian-Chinese laboratory of radiation control and inspection" in cooperation with the Laboratory No. 42 (High Current Betatron) INDT TPU.

Betatron tomography system (Figure 1) consists of a betatron (X-ray source), the mechanical system, the detector, the control systems and the image reconstruction system.

2. Method of Calculation

The quality of a CT depends on several factors, the most important ones are the geometric magnification and spatial resolution. Spatial resolution refers to the ability of CT system to detect small details or contains of small features with respect to the reference point.

There are many factors that influence the spatial resolution of reconstructed images [5]: focal spot size of the x-ray source, detector characteristics, magnification, number of projections, reconstruction algorithms and data processing method. Focal spot size plays a particularly important role in determining image quality and is one of the main factors limiting the spatial resolution [6]. CT systems therefore often classified by the focal spot size, and in our case, the betatron tomograph belongs to the macro-CT (the focal spot size 0.3 mm) [5].

Figure 1. Betatron tomography system.

The calculation formula (1) was first proposed by Yester and Barnes [7] to determine the total unsharpness of the object. As later used to calculate the effective aperture [8], and then used to evaluating the spatial resolution in a given system [9]. We also apply this formula to determining the spatial resolution of betatron tomograph:

$$BW \cong \frac{\sqrt{d^2+[a\cdot(M-1)]^2}}{M} \quad (1)$$

where *BW* is beam width, the major limitation on spatial resolution in a given CT system; *a* is focal spot size; *d* is detector aperture size; *M* is geometric magnification on the axis of rotation. The scanning scheme of a CT system is shown in Figure 2 [9-10], the geometric magnification $M = L/S$ (where *L* is the distance from the radiation source to the detector, *S* is the distance from the radiation source to the axis of rotation).

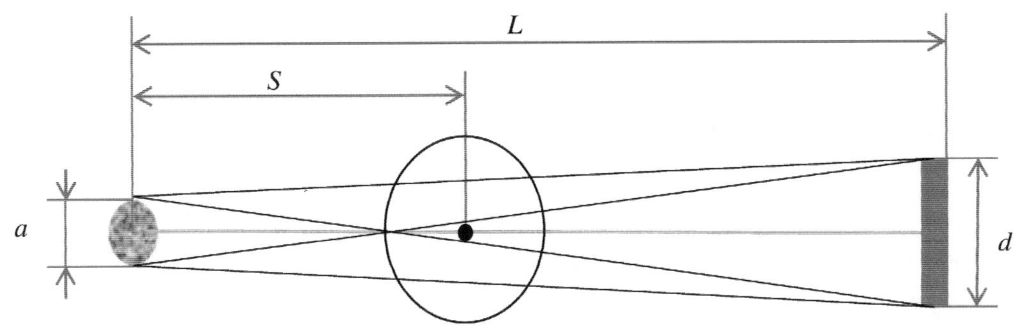

Figure 2. Scanning scheme of CT system.

Detailed mathematical analysis of formula (1) was made by Chen J. [10], in his work, he converted formula (1) to expression (2):

$$BW=\left((a^2+d^2)\cdot\left(\frac{1}{M}-\frac{a^2}{a^2+d^2}\right)^2+\frac{a^2\cdot d^2}{a^2+d^2}\right)^{\frac{1}{2}} \quad (2)$$

It is known that, as one kind of X-ray resource, betatron has the focal spot size of a = 0.3 mm; and for betatron tomograph the detector aperture size d = 0.4 mm. Substituting the values of a and d in expression (2), we obtained the dependence BW of M (3):

$$BW=\left(0.25\cdot\left(\frac{1}{M}-0.36\right)^2+0.058\right)^{\frac{1}{2}},\ [\mathrm{mm}] \quad (3)$$

With the help of Mathcad, we got a plot of the dependence BW of M ($1 < M < 100$), shown in Figure 3.

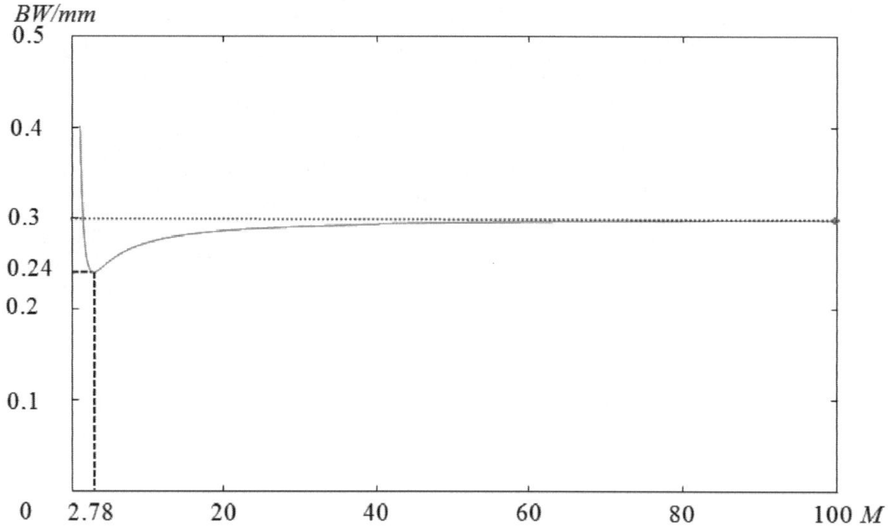

Figure 3. Dependence BW of M with the use of betatron.

With the analysis of dependence BW of M, it is noticeable that with the increase of M, the value of BW decreases rapidly and reaches a minimum value BW = 0.24 mm where M = 2.78. Then, with the increase of M, the value of BW increases and converges to a constant whose value is equal to the size of the focal spot.

3. Result and Discussion

Development of betatron tomography system isn't an easy task, the definition and evaluation of spatial resolution is only one of a number problems. Since the spatial resolution depends on many factors, so in the process of developing the betatron tomograph should find a compromise between these factors to ensure optimal performance.

During the development process of betatron tomograph, we use X-ray tube MXR-451HP/11 (Figure 1) with focal spot size a_1 = 1 mm to make a comparative analysis. Substituting the values of a_1 and d in expression (2), we got the relationship BW from the M (Figure 4).

It is noticeable that with the increase of M, the value of BW decreases rapidly and reaches a minimum value BW = 0.37 mm where M = 1.14. Then, with the increase of M, the value of BW increases and converges to a constant whose value is equal to the size of the focal spot.

Further we analyzed the formula (2) with a mathematical method. We know that M is from 1 to ∞.

When $M \rightarrow 1$, the formula (2) has the form:

$$\lim_{M\rightarrow 1} BW=\left(\left(a^2+d^2\right)\cdot\left(\frac{1}{M}-\frac{a^2}{a^2+d^2}\right)^2+\frac{a^2\cdot d^2}{a^2+d^2}\right)^{\frac{1}{2}}=d.$$

When $M \rightarrow \infty$, the formula (2) has the form:

$$\lim_{M\rightarrow \infty} BW=\left(\left(a^2+d^2\right)\cdot\left(\frac{1}{M}-\frac{a^2}{a^2+d^2}\right)^2+\frac{a^2\cdot d^2}{a^2+d^2}\right)^{\frac{1}{2}}=a.$$

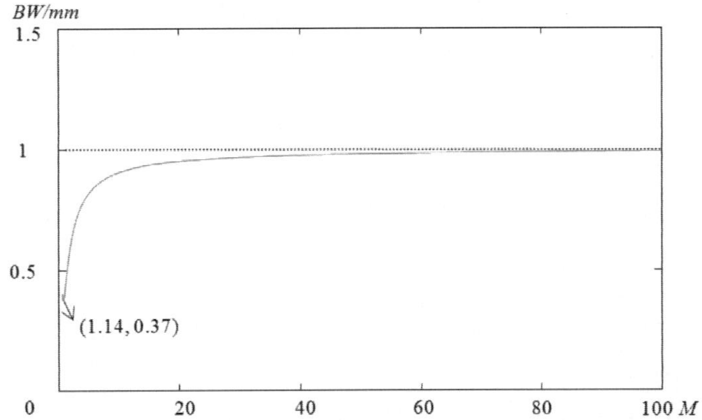

Figure 4. Dependence *BW* of *M* with the use of X-ray tube.

The minimum value of *BW* is obtained by $\frac{1}{M}-\frac{a^2}{a^2+d^2}=0$, which means $M=1+\frac{d^2}{a^2}$. Therefor we got the minimum value $BW=\frac{a^2 \cdot d^2}{a^2+d^2}$.Then we got the dependence *BW* of *M* in Figure 5.

With the analysis of dependence *BW* of *M* in Figure 5, we can see that the initial value of *BW* with $M = 1$ is equal to the detector aperture size *d*. Then with the increase of *M*, the value of *BW* decreases and reaches a minimum value of $BW = \frac{a^2 \cdot d^2}{a^2+d^2}$ mm where $M = 1+\frac{d^2}{a^2}$. After that, with the increase of *M*, the value of *BW* increases and converges to a constant whose value is equal to the size of the focal spot *a*.

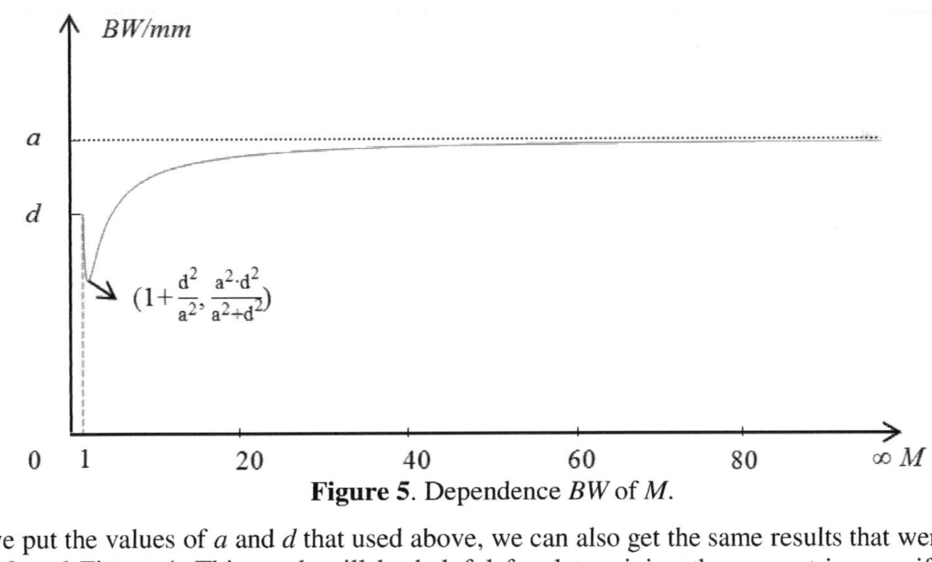

Figure 5. Dependence *BW* of *M*.

If we put the values of *a* and *d* that used above, we can also get the same results that were shown in Figure 3 and Figure 4. This work will be helpful for determining the geometric magnification and spatial resolution in the future. And it also helps to find the most suitable value *M* for a CT system in practice.

4. Conclusions

Since the beam width BW is a major limitation on the spatial resolution of a CT system, it is possible to evaluate the spatial resolution of a CT system indirectly by evaluating the *BW*.

From the analysis of the dependence *BW* of *M*, we can make the following conclusions:

a) $M = 2.78$ is theoretically the optimal magnification for betatron tomograph;

b) optimal spatial resolution can be obtained by $BW = 0.24$ mm;
c) the geometric magnification M, used for betatron tomograph has a value of $1 < M < 20$.

The result of the calculation of $M = 2.78$ enters the experimentally determined values of the interval M: $1.5 < M < 3$.

References

[1] Zhang Ch *et al* 2009 *Industrial CT techniques and principles* (NSW: Science Press)
[2] Moskalev V A *and* Chaklov V L 2009 *Betatrons* (Tomsk: TPU Publish House) (in Russian)
[3] Moskalev V A *and* Sergeev G I 2012 *The Induction Electron Accelerator – Betatron* (Tomsk: TPU Publish House) (in Russian)
[4] Chakhlov S V *et al* 2016 *J. Phys.: Conf. Ser.* **671** 012024 DOI: 10.1088/1742-6596/671/1/012024
[5] Chiffre L-D *et al* 2014 *CIRP Annals - Manufacturing Technology* **63** 655–677
[6] Arabi H *et al* 2010 *Iran. J. Radiat. Res* **8(1)** 37–43
[7] Yester M V *and* Barnes G T 1977 *SPIE* **VI (127)** 296–303
[8] *Specification and acceptance testing of computed tomography scanners* 1993 Retrieved from http://www.aapm.org/pubs/reports/rpt_39.pdf
[9] *ISO 15708-1:2002 Non-destructive Testing – Radiation Methods – Computed tomography* 2002 Retrieved from https://www.iso.org/standard/28106.html
[10] Chen J *et al* 2014 *CT Theory and Applications* **23** 37–43

Estimation of the Lithospheric Component Share in the Earth Natural Pulsed Electromagnetic Field Structure

S Y Malyshkov[1], V F Gordeev[1], V I Polyvach[2], S G Shtalin[3], and K N Pustovalov[2]

[1]Senior Researcher, Institute of Monitoring of Climate and Ecosystems, Siberian Branch of the Russian Academy of Science, Tomsk, Russia
[2]Junior Researcher, Institute of Monitoring of Climate and Ecosystems, Siberian Branch of the Russian Academy of Science, Tomsk, Russia
[3]Leading Electronics, Institute of Monitoring of Climate and Ecosystems, Siberian Branch of the Russian Academy of Science, Tomsk, Russia

E-mail: msergey@imces.ru

Abstract. Article describes the results of the atmosphere and Earth's crust climatic and ecological parameters integrated monitoring. The estimation is made for lithospheric component share in the Earth natural pulsed electromagnetic field structure. To estimate lithospheric component we performed a round-the-clock monitoring of the Earth natural pulsed electromagnetic field background variations at the experiment location and measured the Earth natural pulsed electromagnetic field under electric shields. Natural materials in a natural environment were used for shielding, specifically lakes with varying parameters of water conductivity. Skin effect was used in the experiment – it is the tendency of electromagnetic waves amplitude to decrease with greater depths in the conductor. Atmospheric and lithospheric component the Earth natural pulsed electromagnetic field data recorded on terrain was compared against the recorded data with atmosphere component decayed by an electric shield. In summary we have demonstrated in the experiment that thunderstorm discharge originating electromagnetic field decay corresponds to the decay calculated using Maxwell equations. In the absence of close lightning strikes the ratio of field intensity recorded on terrain to shielded field intensity is inconsistent with the ratio calculated for atmospheric sources, that confirms there is a lithospheric component present to the Earth natural pulsed electromagnetic field.

1. Introduction

Though Earth natural pulsed electromagnetic fields (ENPEMF) are widely used in engineering and geophysical surveys, geodynamic forecasts and permafrost studies [1], the discussion about the nature and sources of those fields is still open in the scientific community.

A large number of researchers think that a vast majority of impulses in ENPEMF structure originates from thunder strikes in the atmosphere [2–4], and the proved efficiency of the methods based on those fields recording is explained by the fact that in very low frequency band electromagnetic waves propagate within the Earth-Ionosphere waveguide. Depending on geological framework and Earth's crust stress-strained state, waves' propagation conditions vary and the depth of waves' penetration into geological substrate varies as well.

Another group of researchers, including the authors of this paper, while not denying the existence of the atmospheric component in ENPEMF, proves that lithospheric component takes considerable place in those fields overall structure [5–8]. Radio-wave methods applied are based on the phenomenon of electromagnetic emission - dielectric materials emissive ability when they are acted on. Electromagnetic emission emerges in the process of charges generation and relaxation on fracture planes during the stress state of the rocks. Pulses emerge both when dielectric uniformity changes and when electrolyte-filled capillars rift [9]. In rock formations there can be the following sources of natural electromagnetic fields: soil structure inconsistencies, unequally strained structures, fractures and microfractures. All those sources generate pulsed electromagnetic fields as a result of mechanic-to-electric energy conversion, strain waves from the mantle, tides, microseisms, winds, and technological loads thus creating a natural electromagnetic background of lithospheric nature. Observing electromagnetic emission allows to monitor stress-strained state of the rock formation [10, 11].

According to everything mentioned above it is necessary to create physical-mathematical models, describing electromagnetic processes and to confirm current understanding with an experiment. We carried out an integrated monitoring of climate and ecological parameters of the atmosphere and Earth's crust to estimate lithosphere component share in ENPEMF structure. That is a fundamental research leading to acquiring new knowledge in Earth sciences, but it has a useful applied potential as well. ENPEMF based geophysical survey methods developed were successfully tested in the inland and proved most informative compared to other cost-effective methods.

Most promising locations for new natural deposit site explorations are continental shelves, especially in Extreme North region. If most of the pulses in the ENPEMF structure originate from the atmosphere then applying ENPEMF methods in shelf exploration would be impossible due to the fact that saltwater's high conductivity would screen atmospheric pulsed fields.

The goal of the experiment is to prove the existence of lithospheric ENPEMF component, show the possibility of geophysical survey based on ENPEMF parameters while working on continental shelf. To achieve that goal a series of experiments were carried out recording ENPEMF at the location shielded from atmospheric discharges. Natural materials in a natural environment were used for shielding, specifically lakes with varying parameters of water conductivity. Time of the experiments in July was selected based on seasonal highest ENPEMF intensity with distinguishing extremums of daily trajectories..

2. Experiment

Field research took place in Khakassia republic, Russia. Lake Itkol is a large freshwater pond of 2300 hectares in area. Lake Shira is large water body with mineralization parameters close to seawater. Lake Tus is a bitter lake with a large bottom sludge layer.

Electric field intensity measurement device "CS110" allowing to register a vertical component of electric field intensity in a band of ±22.3kW/m was used to analyze thunderstorm factor. By the operating concept that device is a field mill with a shuttling screening plate.

ENPEMF variation were measured by geophysical recorders. "GR-01" records the following parameters of ENPEMF:

- number of field's magnetic component pulses in a specified time frame, with a pulse height over a specified threshold;
- pulse height of the first pulse exceeding a threshold for magnetic component;

Measurement resolution was 1 second. Meteorological data obtained was compared against the data of meteorological stations network. The closest meteorological station was Shira, situated in about 10km from the experiment site.

We have carried out a research on the particulars of atmospheric-electric and meteorological parameters changes during the passing of the frontal and air-mass convective clouds over the experiment site. To interpret the data, an additional analysis of synoptical maps, nephanalysis and cloud evolution forecast maps and MODIS (KA Terra, Aqua) spectroradiometer data was used including resulting thermodynamical and optical cloud parameters.

To estimate lithospheric component we performed a round-the-clock monitoring of ENPEMF background variations at the experiment location and measured ENPEMF under electric shields. Terrain experiment site registering ENPEMF background variations, electric field intensity and meteorological parameters was located at north-east coastline of Lake Itkol. Recorders in hermetically sealed radio transparent containers were doused to different depths into lakes with different water conductivity (l. Itkol, l. Shira, l. Tus) in order to shield off atmospheric electricity. Skin effect was used in the experiment – it is the tendency of electromagnetic waves amplitude to decrease with greater depths in the conductor. Atmospheric and lithospheric component ENPEMF data recorded on terrain was compared against the recorded data with atmosphere component decayed by an electric shield. Table 1 shows experimentally measured electrical conductivity, skin-layer for that conductivity and calculated based on Maxwell equations atmosphere EM field decaying for different depths. Skin depth is a level inside of the conductor decaying EM field *e* times.

Table 1. Calculated decay of atmospheric EM field depending on conductivity and dousing depth.

	l. Itkol	l. Shira	l. Tus
Conductivity	0.065 [S/m]	1.85 [S/m]	9.8 [S/m]
Skin depth for f=14.5 KHz, [m]	11.6 [m]	2.2 [m]	0.9 [м]
Decay at 2m, [times]	1.2	2.4	8.3
Decay at 10m, [times]	2.4	99.7	-
Decay at 13m, [times]	2.8	396.6	-

Figure 1 shows ENPEMF recorded data in two dominating directions of reception on terrain and atmospheric electric field intensity.

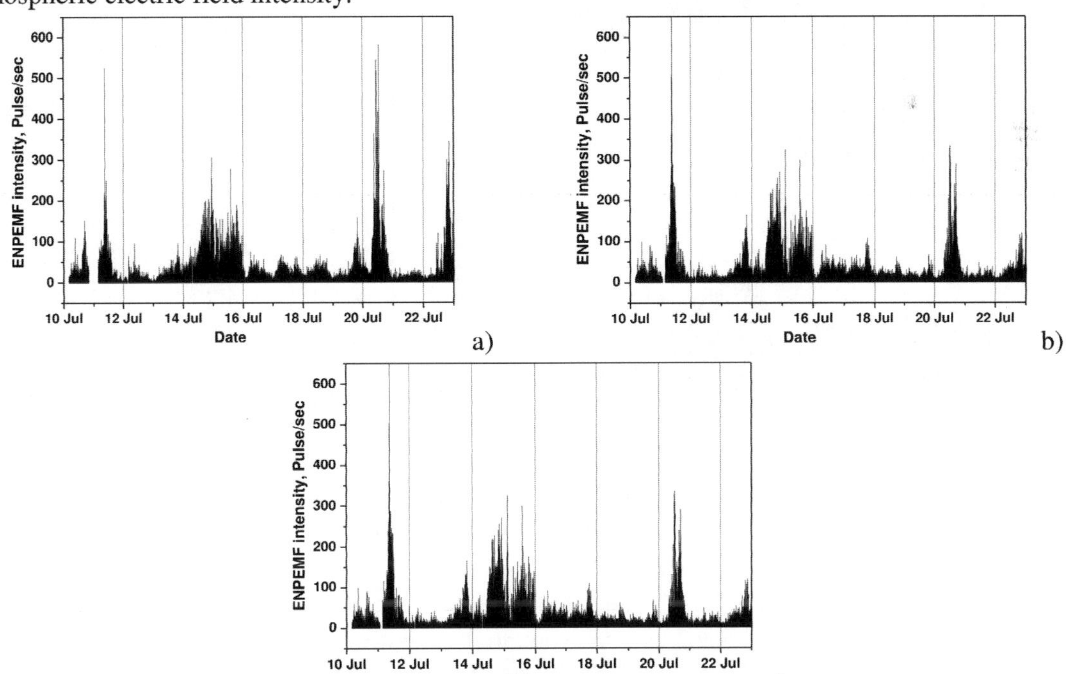

Figure 1. EM field parameters: a) ENPEMF pulse flow intensity in dominating reception direction North-South; b) ENPEMF pulse flow intensity in dominating reception direction West-East; c) Atmospheric electric field intensity.

It is clearly visible that ENPEMF intensity recorded in North-South direction correlates well with the intensity recorded in West-East direction. Anomalies appear on the electric field intensity chart when a thunderstorm cloud was passing over the recording equipment with high peaks during thunder strike discharges. It can be concluded based on Figure 1 that nearby thunderstorm manifest themselves in higher ENPEMF intensity values which is correct.

Let's review the ENPEMF recorded data with non-anomaly electric field intensity values compared to general synoptic situation at the region, e. g. July the 10th and July the 17th.

There was a cold frontal passage over the recording site at July the 10th. Many meteorological station indicated thunderstorm along the front line visually (red circles on the map in Figure. 2b), including the stations within 300 km distance of the ENPEMF recording site. Figure 2a shows ENPEMF intensity values for that day. Red line on the chart illustrates hourly running average value. Running average values do not exceed 5 pulses per second throughout the whole day.

On the July the 17th none of the meteorological stations located within 400 km distance from the experiment site detected thunderstorms. The region was under the anticyclone influence. Despite the absence of nearby thunder strikes running average ENPEMF values exceed those recorded during the thunderstorm passage for over 4 times. That can be considered as one of the indirect proofs to the inconsistency of the theory stating atmospheric component predominance in ENPEMF structure.

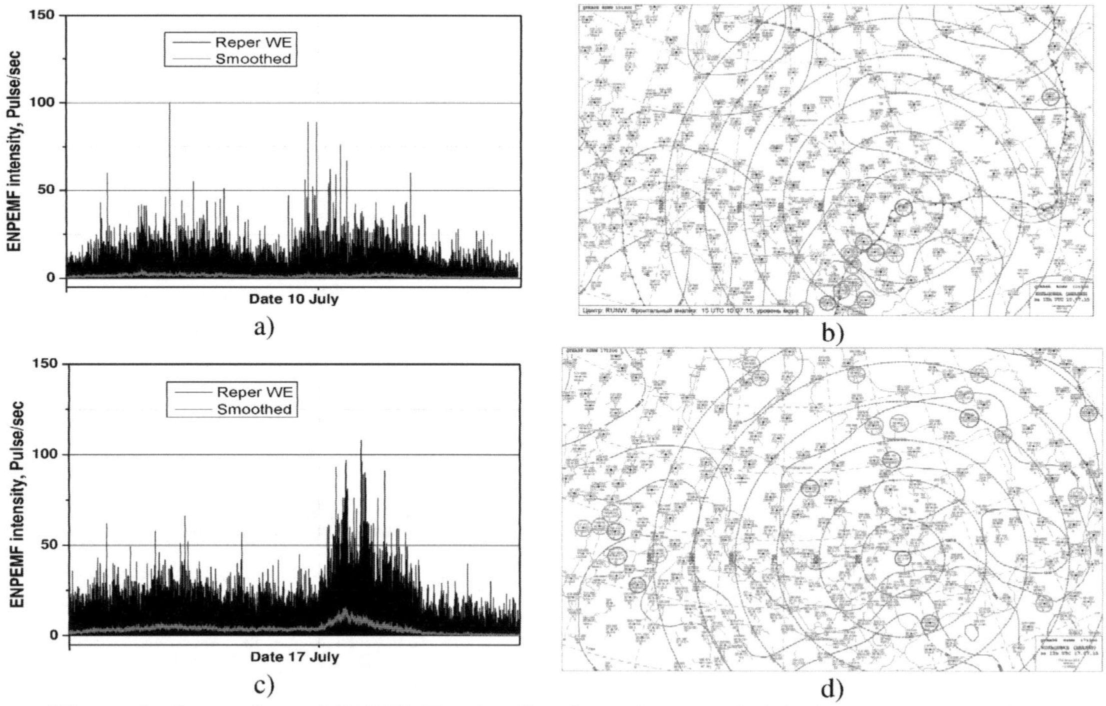

Figure 2. Comparison of ENPEMF pulse flow intensity recorded during atmospheric frontal passage at July the 10th (a) against intensity during anticyclone at July the 17th b) synoptic map at July the 10th; d) synoptic map at July the 17th.

Figure 3 shows ENPEMF intensity values, measured by two recorders one of which was on the bank terrain and another one 13 m underwater of lake Itkol. For better visibility underwater measured values are multiplied by -1. The gap in the lake Itkol underwater measurements is a result of thunderstorm weather floating craft watering ban. Battery elements in the equipment weren't changed in time and it caused a stop in recording. Based on the calculated values for water conductivity of 0.065 cm/m in the case of only atmospheric ENPEMF component present, the field decay recorded underwater should be 2.8 times. However throughout the whole measurements time the observed experimental value was 1.6 times, which confirms the hypothesis of a lithospheric ENPEMF component present. Decay value close to a calculated one was recorded one time during a strong thunderstorm at July the 11th. There was a frontal passage of local cumulonimbus cluster, developing at a rear of a cyclone behind a katafront. According to MODIS cumulonimbus at that cluster developed at a limited local area but had a large size and vertical thickness.

Lets review EM parameters during a thunderstorm of July the 11th in more detail. Figure 4a) shows a segment of ENPEMF recording corresponding to that event.

For better visibility there are EM field intensity values measured close to terrain geophysical recorder overlaid on the chart. Sharp peaks of intensity are caused by a thunderstorm. Sharp peaks in intensity are induced by the close lightning strikes. Plot indicates that ENPEMF atmospheric component decay by the water layer corresponds with a calculated one, but between the lightning strikes there are impulses that are not decayed. That is a lithospheric component. At that time there was a passing of local cumulonimbus cluster, developing at a rear of a cyclone behind a katafront. One of the clouds of that cluster passing over the experiment site, should have reach a mature stage, as indicated by high values of EM field intensity near the surface and intense thunderstorm activity (peaks in EM field intensity at Figure 4a), and anvil form observed at passing.

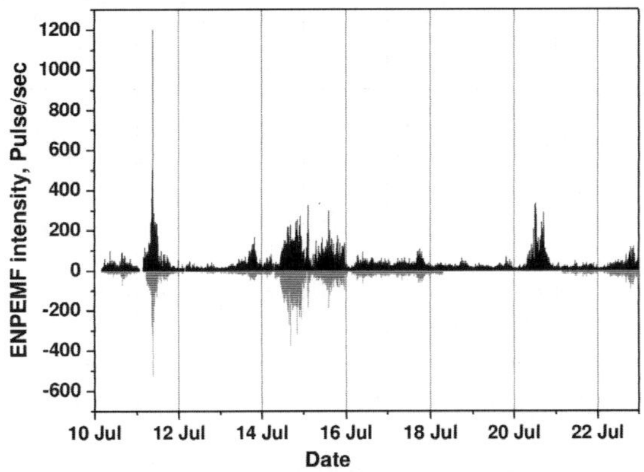

Figure 3. ENPEMF intensity on terrain and 13 m underwater of lake Itkol. Underwater values multiplied by - 1.

Figure 4b shows segments of ENPEMF recorded on terrain and 13 m underwater of lake Itkol after the thunderstorm. It follows from the charts that in the absence of nearby lightning strikes there is no decay to ENPEMF intensity by water layer electric shield neither considering pulse flow intensity nor pulse amplitude. That is another proof of the lithospheric component existence in those fields structure.

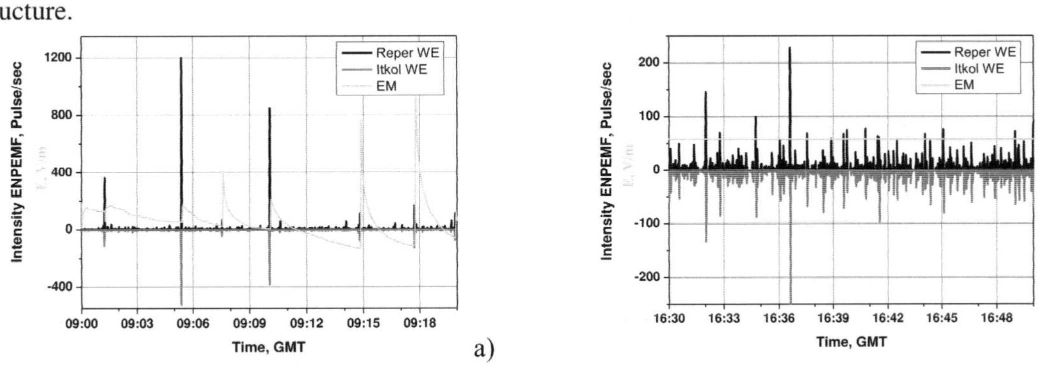

Figure 4. A segment of ENPEMF intensity recording and EM field intensity recording. Underwater intensity values multiplied by – 1, where a) during a thunderstorm; b) in the absence of thunderstorms.

In addition to the results of ENPEMF measurements in a freshwater lake presented above, there was also a simultaneous fields recording done underwater of salted lake Shira and on terrain, underwater of bitter lake Tus and on terrain. On lakes Shira and Tus at 2m underwater the decay value was 1.5 times higher than a calculated value. Those lakes have a sludge layers with mineralization higher than water mineralization, so the equipment was in fact shielded from both atmospheric and lithospheric electricity. Experiment program didn't include sludge layers' depth and conductivity; therefore it wasn't possible to calculate the decay for those conditions using known formulae. At 13 m underwater of lake Shira the decay recorded was 1380 times while calculated decay was 380 times for conductivity of 1.85 S/m. According to Malakhov's plan of sludge deposit the thickness of deposits at measurement site is around 1m

3. Summary

In summary we have demonstrated in the experiment that thunderstorm discharge originating electromagnetic field decay corresponds to the decay calculated using Maxwell equations. In the absence of close lightning strikes the ratio of field intensity recorded on terrain to shielded field intensity is inconsistent with the ratio calculated for atmospheric sources; that confirms there is a lithospheric component present to ENPEMF. Bottom sludge deposits of high conductivity apparently set practical limitation to ENPEMF method applications for geophysical survey on inland shelves.

References

[1] Malyshkov Yu P *et al* 2011 RF Patent 2414726 *Bull. Izobret.* **8** (in Russian)
[2] Kozlov V I *et al* 2010 *Geomagnetism and Aeronomy* **49**(8) 1296–1298 DOI: 10.1134/S0016793209080544
[3] Druzhin G I *et al* 2010 *Geomagnetism and Aeronomy* **49(8)** 1305–1307 DOI: 10.1134/S001679320908057X
[4] Druzhin G I *et al* 2016 *Bulletin KRASEC. Physical & Mathematical Sciences* **4(15)** 99–105 (in Russian)
[5] Vorobyov A A 1970 *Geology and Geophysics* **12** 3–13 (in Russian)
[6] Malyshkov Yu P *and* Malyshkov S Yu 2012 *Eccentric rotation of the earth's core and lithosphere: Origin of deformation waves and their practical application* (Taiwan: Nova Science Publishers) **6** 115–212
[7] Surzhikov A P *et al* 2012 *Therm. Anal. Calorim.* **108(3)** 1207–1212 DOI: 10.1007/s10973-011-1734-z
[8] Gordeev V F *et al* 2016 *IOP Conf. Ser.: Earth Environ. Sci.* **48** 012026 DOI: 10.1088/1755-1315/48/1/012026
[9] Surkov V V 2000 *Electromagnetic effects in case of earthquakes and blasts (*Moscow: MIFI) (in Russian)
[10] Malyshkov Yu P et al 2013 *Earth's Natural Electromagnetic Noises in a Very-Low Frequency Band* (New York: Nova Science Publishers) 1–86
[11] Malyshkov Yu P *and* Malyshkov S Yu 2009 *Russian Geology and Geophysics* **50(2)** 115–130 DOI: 10.1016/j.rgg.2008.06.019

IOP Conf. Series: Materials Science and Engineering **189** (2017) 012024 doi:10.1088/1757-899X/189/1/012024

The Selection of Computed Tomography Scanning Schemes for Lengthy Symmetric Objects

V B Trinh[1], Y Zhong[1], and S P Osipov[2]

[1]Postgraduate, National Research Tomsk Polytechnic University, Tomsk, Russia
[2]Leading Scientific Employee, National Research Tomsk Polytechnic University, Tomsk, Russia

E-mail: vanbac1008@gmail.com

Abstract. The article describes the basic computed tomography scan schemes for lengthy symmetric objects: continuous (discrete) rotation with a discrete linear movement; continuous (discrete) rotation with discrete linear movement to acquire 2D projection; continuous (discrete) linear movement with discrete rotation to acquire one–dimensional projection and continuous (discrete) rotation to acquire of 2D projection. The general method to calculate the scanning time is discussed in detail. It should be extracted the comparison principle to select a scanning scheme. This is because data are the same for all scanning schemes: the maximum energy of the X–ray radiation; the power of X–ray radiation source; the angle of the X-ray cone beam; the transverse dimension of a single detector; specified resolution and the maximum time, which is need to form one point of the original image and complies the number of registered photons). It demonstrates the possibilities of the above proposed method to compare the scanning schemes. Scanning object was a cylindrical object with the mass thickness is 4 g/cm^2, the effective atomic number is 15 and length is 1300 mm. It analyzes data of scanning time and concludes about the efficiency of scanning schemes. It examines the productivity of all schemes and selects the effective one.

1. Introduction

The manufacturing of modern nuclear fuel (NF), according to a given level of functionality, safety and reliability, is impossible without the complex control of NF's parameters. Since the middle of 80s X–ray computed tomography method (CT scan) is used to control the fuel elements [1–4]. CT scan allows valuate the internal object structure, correctness of assembly, etc. Due to the development of computer facilities and X–ray detection systems recently the 3D CT [5–11] is developed rapidly, which differs from the 2D CT scan realization by higher performance. A change of the fuel elements design leads to the necessity of complex theoretical and experimental studies, analysis of the results which will allow make a conclusion about the possibility of CT scan usage for multi–parameter examination of the fuel elements.

2. Basic CT Scan Schemes and Method for Calculation of Scanning Time

Let us discuss all the most popular scanning schemes of lengthy symmetric objects, a cylindrical fuel element with diameter H and length L is presented as example. The performance is inversely proportional to the total examination time T of the object. For all schemes the time T is defined by t_0 is the residence time of the detector systems in one position (the time to produce a projection), Δt is the time to travel the source–detectors system from one position to another, $\Delta\theta$ is a step of the rotation angle, k is the required number of time intervals for the "black" and "white" calibrations.

Scheme 1: Figure 1 displays the scheme for the classical CT scan method. In this scheme the object rotates discretely or continuously around the axis and a set of one–dimensional projections are acquired. After acquisition of projections for a one layer, the object is moved with scanning step Δx and the process repeats. Scanning step should not exceed the size of the sensitive volume of the detector in the scanning direction d.

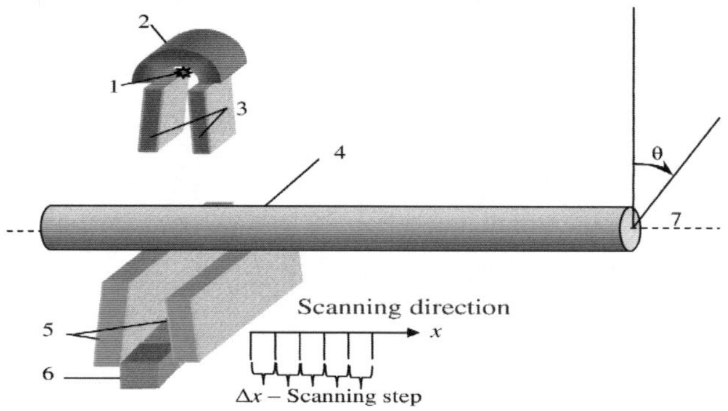

Figure 1. Continuous (discrete) rotation with a discrete linear movement, there 1 is radiation source; 2 is protection of the radiation source; 3 is collimator of source; 4 is scanning object; 5 is collimator of detector; 6 are detectors; 7 is rotation axis of scanning object.

The Equation to calculate the scanning time T for the first scheme has a form (1):

$$T = t_0 \times \frac{180}{\Delta\theta} \times \frac{L}{\Delta x} + \Delta t \frac{L}{\Delta x} + t_0 \times k, \tag{1}$$

Scheme 2: This modification of first scheme is based on the usage of the X–rays panel detector or the matrix of radiometric detectors. The scanning process generates a set of two–dimensional projections (Figure 2). In this scheme the object rotates discretely or continuously around the axis and a set of two–dimensional projections are acquired. After acquisition of projections for some layers, the object is moved with scanning step ΔX and the process repeats. Scanning step should not exceed the size of the sensitive size of the panel detector in the scanning direction D.

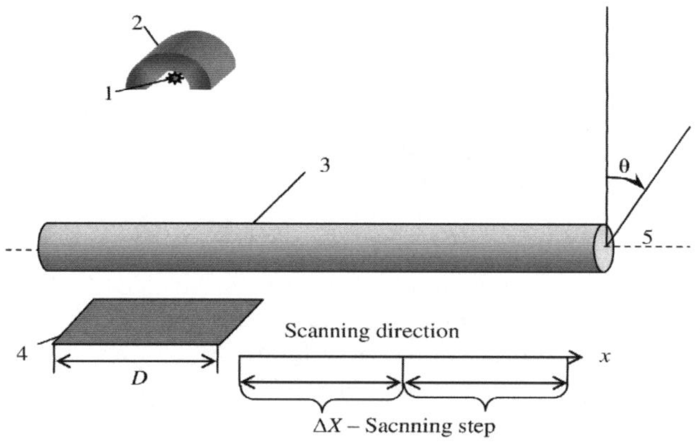

Figure 2. Continuous (discrete) rotation with discrete linear movement to acquire 2D projection, there 1 is radiation source; 2 is protection of radiation source; 3 is scanning object; 4 are matrix of detectors; 5 is rotation axis of scanning object.

The scanning time T is estimated by the Equation (2):

$$T = t_0 \times \frac{180}{\Delta\theta} \times \frac{L}{\Delta X} + \Delta t \frac{L}{\Delta X} + t_0 \times k \,. \tag{2}$$

The difference of second scheme from the first one is the acquisition of 2D projections and a discrete step of linear movement ΔX which is not compared with the size of a single detector d, but it is compared with the size of the detection system in the scanning direction D ($\Delta X \leq D$).

Scheme 3: In first scheme the axis of radiometric detector is perpendicular to the rotation axis. In this scheme, we use the lengthy radiometric detectors, which overlap a shadow of scanning object allow orientate the axis of radiometric detector parallel to the rotation axis (Figure 3). In this scheme the object moves discretely or continuously perpendicularly the axis and a set of one–dimensional projections are acquired. After acquisition of projections for some layers, the object is rotated with step $\Delta\theta$ and the process repeats.

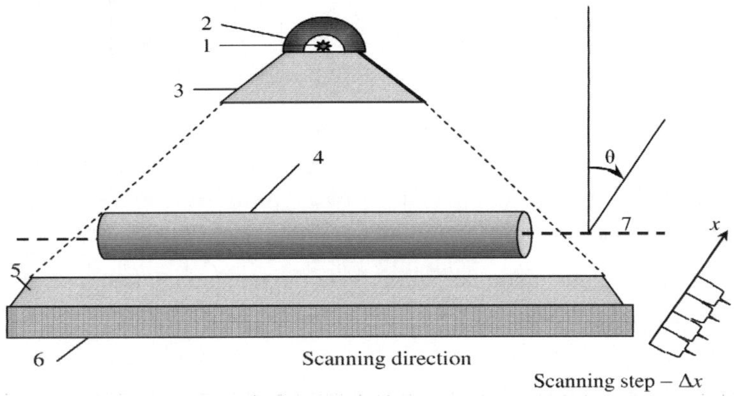

Figure 3. Continuous (discrete) linear movement with discrete rotation to acquire one– dimensional projection, there 1 is radiation source; 2 is protection of radiation source; 3 is collimator of source; 4 is scanning object; 5 is collimator of detector; 6 are detectors; 7 is rotation axis of scanning object.

As result of the discrete linear movement of scanning object relatively to the radiation source–detectors system, the first stage is a formation of shadow image. Then object is rotated on a predetermined angle and operation to form the one–dimensional projection is repeated as necessary.

Equation to evaluate the scanning time T for scheme 3 is (3):

$$T = t_0 \times \frac{180}{\Delta\theta} \times \frac{D_0}{\Delta x} + \Delta t \frac{180}{\Delta\theta} + t_0 \times k \,, \tag{3}$$

Here D_0 is transverse size of the scanning object shadow.

Scheme 4: The following scheme differs from the third scheme in X–ray recorder, as which is a panel detector or matrix of radiometric detectors. The mandatory requirement for the size of the detection system: The shadow of the scanning object is placed completely on the frontal surface of the detection system. Compliance with this requirement excludes the linear movement of the object, and retains only the rotation (Figure 4). In one position, two–dimensional projections of the whole object are acquired. After that, the object is rotated with step $\Delta\theta$ and the process repeats.

The Equation to calculate the scanning time T is:

$$T = t_0 \times \frac{180}{\Delta\theta} + \Delta t \frac{180}{\Delta\theta} + t_0 \times k \,, \tag{4}$$

There are many other scanning schemes, but all of them are derived from the schemes, which are shown in Figures 1–4.

For each of scanning schemes we must calculate the total time T, which need to receipt information to reconstruct a 3D structure.

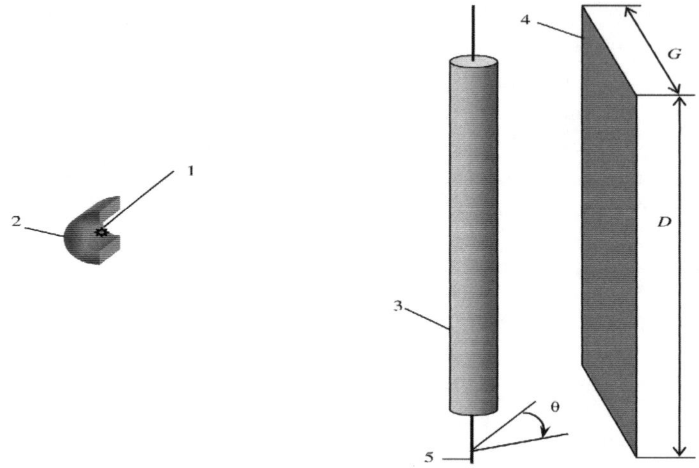

Figure 4. Continuous (discrete) rotation to acquire of 2D projection, there 1 is radiation source; 2 is protection of radiation source; 3 is scanning object; 4 are matrix of detectors; 5 is rotation axis of scanning object.

3. The Comparative Parameters of Scanning Scheme Productivity

The comparison principle to select a scanning scheme can be expressed by the phrase – "ceteris paribus". For "ceteris paribus" in this case is related that the data are same for all scanning schemes. The initial data are: $E_{\max}$ is the maximum energy of the X–ray radiation; P is a power of X–ray radiation source; ω is an angle of the X-ray cone beam; d is the transverse dimension of a single detector; r_{lim} is specified resolution, H is the diameter of the object.

Another parameter must have the same value for all scanning schemes, it is the maximum time t_0, which is need to form one point of the original image and complies the number of registered photons n_0. Time t_0 is chosen to maximize a value of product of the geometric easing and radiation easing. We assume that the anisotropy of X–ray radiation is insignificantly, the Equation to evaluate t_0 has the form (5):

$$t_0 = cP(\xi,\varphi)\max_{\xi,\varphi}\frac{e^{-\mu(E_{eff}(\rho H(\xi,\varphi)))\rho H(\xi,\varphi)}}{F^2(\xi,\varphi)}, \tag{5}$$

Here ξ, φ are the polar and azimuth angles in the coordinate system, which is associated with the radiation source; $\rho H(\xi,\varphi)$ is the average mass thickness of the object in the direction ξ, φ; E_{eff} is effective energy of X–rays; c is recalculated coefficient of power in the number of particles. In the first and third schemes only polar angle is significant.

The above presented method is allowed to compare productivity of different scanning schemes for lengthy symmetric objects.

4. Example of Calculation for ScanningTime

Let demonstrate the possibilities of the above proposed method to compare the scanning schemes. Scanning object was the cylindrical object. The mass thickness is 4 g/cm^2, the effective atomic number is 15 and length is 1300 mm. The transverse size of a single detector is 0.1×0.1 mm^2. Angle θ = 40°, 60°, 80°. Highest energy of radiation was $E_{max}=250\ keV$.

We calculated the scanning time for all schemes. The results of calculation are summarized in Table 1.

Table 1. Scanning time *T*.

Angle θ, °	Scanning scheme	Scanning time (unit of time)
40	1	1.000
40	2	0.840
40	3	30.000
40	4	1.459
60	1	1.000
60	2	0.039
60	3	20.570
60	4	0.665
80	1	1.000
80	2	0.019
80	3	13.419
80	4	0.328

5. Conclusions

From the analysis data we can conclude that the scanning scheme "Continuous (discrete) rotation with discrete linear movement for the formation of 2D projection" has greatest efficiency. The scheme "Continuous (discrete) linear movement with discrete rotation for the formation of one–dimensional projection" is least effective. The reason of this fact is the necessity to save the radiation field. That leads to increase in distance from the radiation source to the scanning object. With increasing the scanning angle, the efficiency of the scanning schemes is changed and scheme "Continuous (discrete) linear movement with discrete rotation for the formation of one–dimensional projection" would be the most inefficient. The selection between the scanning scheme "Continuous (discrete) rotation with discrete linear movement for the formation of 2D projection" and scheme"Continuous (discrete) rotation for the formation of 2D projection"must do the user.

References

[1] Kosarev L I *et al* 1987 *Atomic Energy* **62** 14–19 DOI: 10.1007/BF01127408
[2] Zhukov Yu A *et al* 1997 *Avtometriya* **4** 43–50 (in Russian)
[3] Kosarev L I *et al* 2006 *Zavodskaya Laboratoriya* **72(1)** 32–35 (in Russian)
[4] Gras Ch *and* Stanley S J 2008 *Annals of Nuclear Energy* **35(5)** 829–837 DOI: 10.1016/j.anucene.2007.09.017
[5] Yang M *et al* 2014 *Energy* **68** 385–398 DOI: 10.1016/j.energy.2014.02.076
[6] Caruso S *and* Jatuff F 2014 *Progress in Nuclear Energy* **72** 49–54 DOI: 10.1016/j.pnucene.2013.09.007
[7] Parker H O *and* Joyce M J 2015 *Progress in Nuclear Energy* **85** 297–318 DOI: 10.1016/j.pnucene.2015.06.006
[8] Ishimi A *et al* 2015 *Nuclear Technology* **189(3)** 312–317 DOI: 10.13182/NT14-34
[9] Plotnikova I *et al* 2016 *IOP Conf. Ser.: Mater. Sci. Eng.* **132** DOI: 10.1088/1757-899X/132/1/012023
[10] Hatch G M. *et al* 2014 *Journal of Forensic Radiology and Imaging* **2(2)** 52–59 DOI: 10.1016/j.jofri.2014.02.039
[11] Baglivo M *et al* 2013 *Journal of Forensic Radiology and Imaging* **1 (1)** 3–9 DOI: 10.1016/j.jofri.2012.10.003

The Level of Impact Noise and Threshold Characteristics Nonlinear Assessment Signal Parameters

Y S Timoshenkova[1]

[1]Student, Ural Federal University named after the First President of Russia B.N.Yeltsin, Yekaterinburg, Russia

E-mail: juliatimoshenkova@gmail.com

Abstract. The article deals with the effect of noise which influences the system with simple signal formation, the output characteristics of the system. The system is further subject to nonlinear signal transformation. We received results and carried out analysis of signal/noise ratio at the output of the system with linear and non-linear release of the information signal parameters, such as amplitude, square amplitude, phase and double phase. We studied the effect of noise bandwidth of the linear part of the system and the noise bandwidth of the system after the nonlinear transformation on the signal/noise ratio at the output of the system. We analyzed thresholds valuations of signal/noise ratio at the input of the nonlinear part, limiting the area of noise signal suppression at its output.

1. Introduction

Evaluation of information parameters of signal such as amplitude and phase of signal are usually produced using methods of non-linear transformations. As a result, there is a change of the signal/noise ratio (SNR) of the process, at the output of nonlinear device. Character of this process often depends on the type of non-linear transformations. In the paper we analyze the effect of noise at the input of the system on change the SNR in the nonlinear transformation of various types of signal and threshold characteristics. The purpose of the research is to analyze changes in the SNR with the non-linear transformation of signal.

2. Experimental

Mathematical description of the processes in nonlinear systems, including analysis of the interaction between the signal and noise; and the calculation of the ratio of their capacities, is a very difficult task, which is usually solved by numerical methods or methods of mathematical modeling. In the work mathematical modeling methods have been used. For the research were selected narrowband partially correlated narrowband processes. The signal was obtained by submitting blank noise on shaping filter. Further blank noise, which simulates the effects of noise on the system, is added. The resulting mixture of signal and noise is supplied to the non-linear converter (Figure 1). There was selected standard polynomial nonlinear conversion power in the research of the following form (1):

$$i = a_0 + a_1 \cdot U + a_2 \cdot U^2 + ... + a_n \cdot U^n, \quad (1)$$

where i is received signal, U is the signal amplitude, a is approximating coefficients.

It is assumed that the information lies in the fluctuation component of the process at the nonlinear element output, where the correlator that calculates the centered process step autocorrelation function (ACF) with much less than the information process interval correlation, is used to separate that component. In this case the ACF values at zero shift is equal to the total signal power and interference [2–4].

When a shift is not zero, the ACF values at a sufficiently large averaging interval are determined only by the signal because the noise component value at divergent moments of time are uncorrelated and make it possible to estimate the information signal power.

ACF calculations were done using Equations 2–5. Formulas allow us to calculate information parameters of signal such as amplitude, double amplitude, phase and double phase. Represented as set of samples of instantaneous values of the values of sine and cosine component signal.

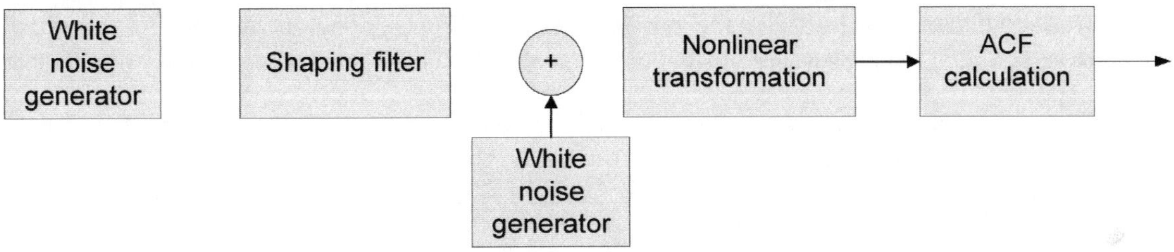

Figure 1. Studied model block diagram.

As an example Figure 2 shows the ACF process at the output of the nonlinear element that provides a signal amplitude assessment according to the Equation [5]:

$$U_i = \sqrt{Uc_i^2 + Us_i^2}\,, \tag{2}$$

where U_i is signal amplitude counting, Uc_i is the instantaneous value of the cosine component signal counting, Us_i is the instantaneous value of the sine component signal counting;

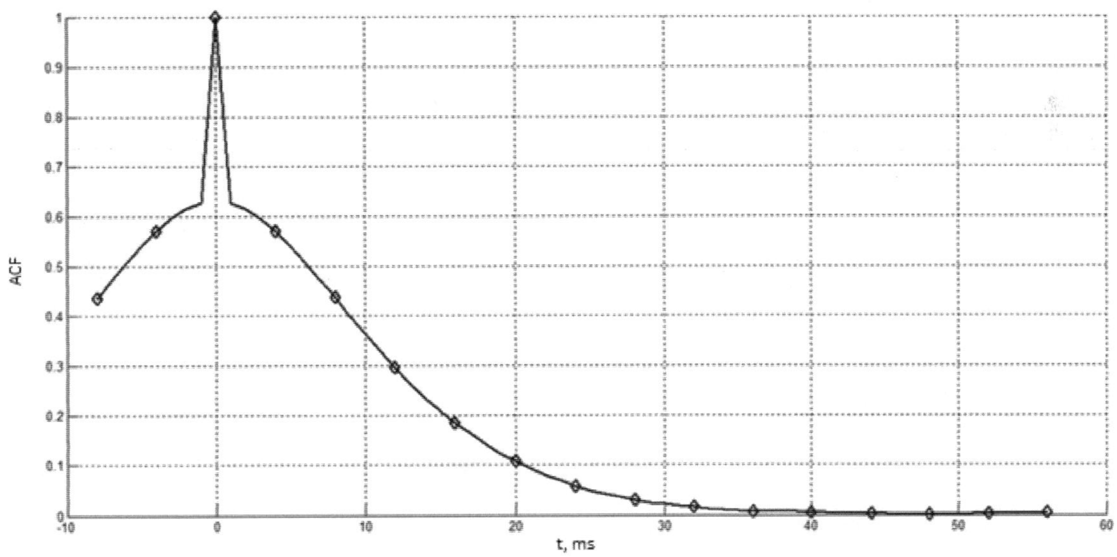

Figure 2. Type of the investigated signal autocorrelation doubled phase.

3. Results and considerations

Four signal characteristics have been taken as information system parameters: the amplitude, square amplitude, phase and double phase, with different signal-to- noise ratio of the system. Besides of the amplitude as an information parameter we consider: squared amplitude, the cosine of the signal phase and the cosine of the doubled phase, calculated according to the Equations (3)–(5) [1–3]:

$$U2_i = Uc_i^2 + Us_i^2, \tag{3}$$

$$C\varphi = \frac{Uc_i}{\sqrt{Uc_i^2 + Us_i^2}}, \tag{4}$$

$$C2\varphi = \frac{Uc_i^2 - Us_i^2}{Uc_i^2 + Us_i^2}. \tag{5}$$

Comparative analysis ACF was performed for various parameters of the signal in the absence of noise and the presence of noise.

As might be expected, in the absence of the noise component in the system, cross-correlation functions (correlation function of sine and cosine component correlation functions sine and cosine phases) are zero, and in the presence of noise there is a correlation between the quadrature components of the signal, which depends on its size.

By the results of the experiment, it was determined that according to the ACF figure we can measure the influence of noise on the system. According to the peak of the ACF we calculate value of the power signal and the amount of noise. At the level of the base of the peak we calculate signal strength. As a result, it is easy to make calculation of the noise power affecting the system as a difference between total power and signal strength (Figure 3).

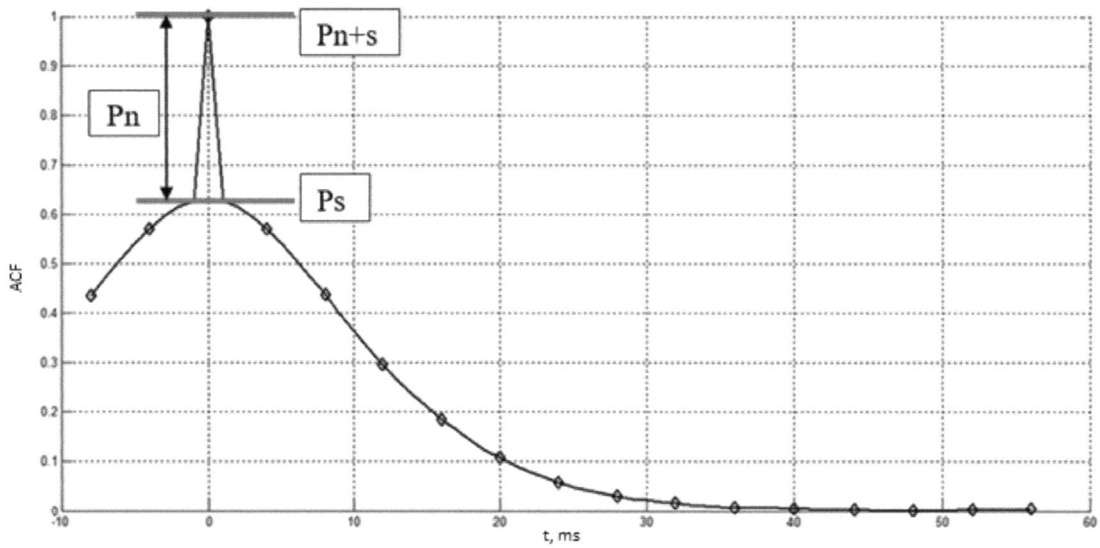

Figure 3. The method of calculating noise power.

On the basis of the realized system, we studied dependence of the SNR at the system output and the SNR at the input of the nonlinear transducer. Plots at Figure 3 show the dependence of the SNR at the output of the nonlinear transducer upom the SNR at the input for various information signal parameters.

The research was conducted by comparing the experimentally obtained data graphically, comparing the mean square error and relative standard error. Figure 4 shows graphs of the SNR at the output of the nonlinear transformer SNR at the input of various information parameters of the signal.

Particular attention should be paid to the range where the Signal-to- Noise Ratio input value does not exceed 10÷20. The graph at Figure 3 is depicting the dependence of the SNR at the output and input of the system for the specified interval.

From these graphs we can see that the smallest decrease in Signal-to- Noise Ratio occurs with using the amplitude and phase, in these cases the average values are not declined by more than 2–3 times. Whereas as the use of a squared amplitude and doubled phase gives a ratio decrease to 4–8 times. The SNR change can be called a system loss. For the amplitude and phase information parameters this values are the smallest, and for the squared amplitude and double phase they are considerably bigger. It is noteworthy that when the SNR value at the input is less than 2, the system loss increases significantly.

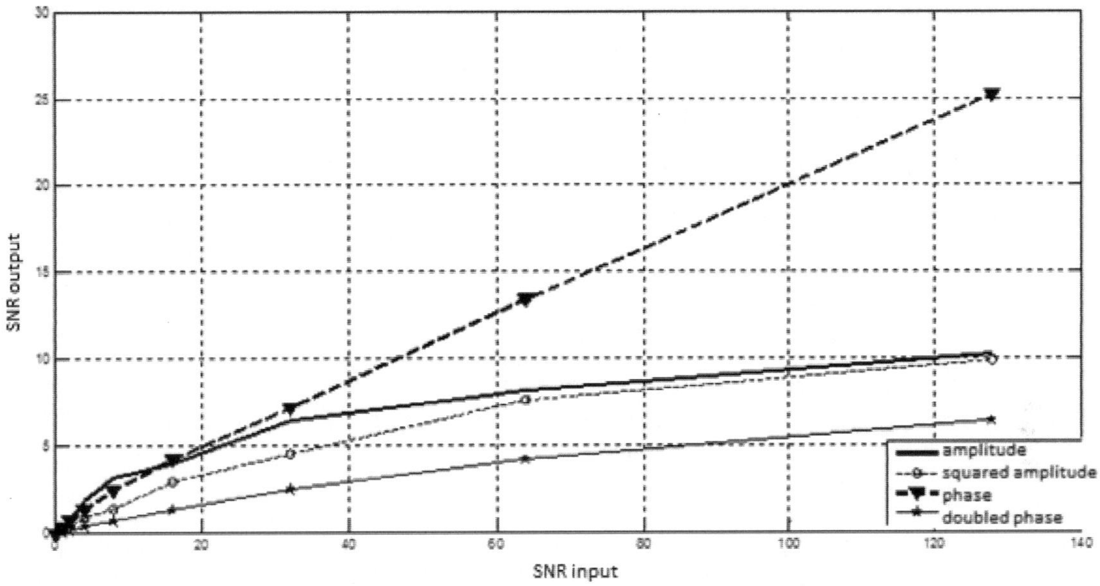

Figure 4. The dependence of the SNR at the input and output of the nonlinear transducer.

Phase signal received as an information signal parameter in the modeling of systems with non-linear signal processing is a promising information parameter in correlation velocity meter, because it does not impose any restrictions on the choice of an automatic gain control parameters and signal strength, and has the smallest width of CCF.

As a consequence, the use phase of the signal as an information parameter of signal allows you to avoid errors in the processing of the signals received at modeling reflected by extended surface. It is worth to note that the dependence of the SNR on such information parameter, as the signal phase, is linear. This fact can be used in the processing of information and simplification of correlation gauges.

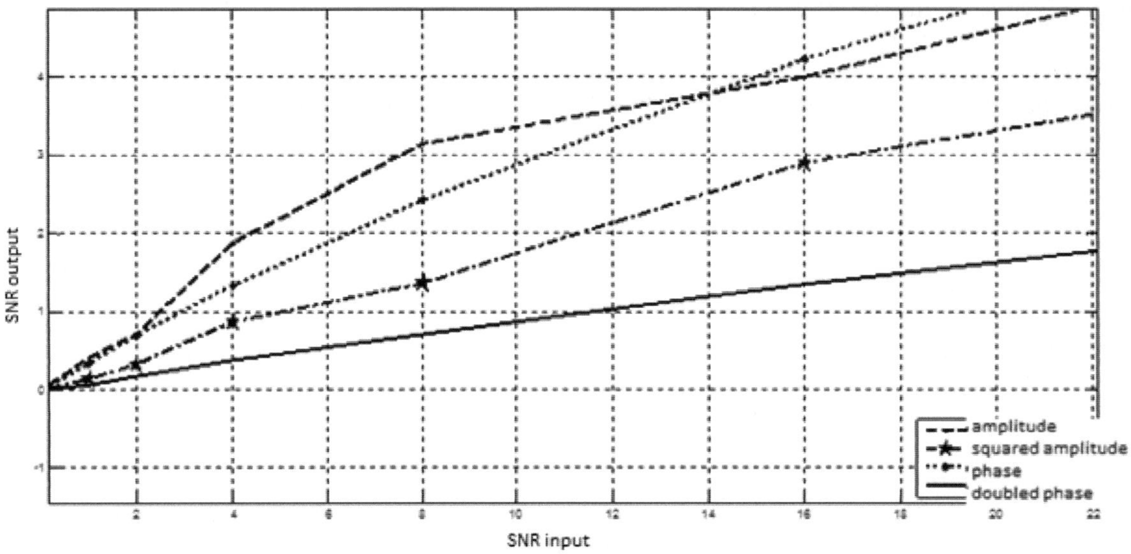

Figure 5. The dependence of the SNR at the system output from the SNR on the system input for the critical interval limited 20.

4. Summary

These results must be considered when modeling systems with signals reflected by extended surface and correlation measurement of parameters (speed, altitude, etc.). When modeling systems with reflected signals, you need to take into account the effect of noise; that's why when using non-linear signal processing methods you should provide at least 6 SNR at the input of such systems.

Increasing the level of interference, determined by random dispersion of the input signal, reduces the useful signal at the output of the nonlinear element. The threshold effect for the considered values of the SNR at the input are not observed.

Increasing the value of the SNR at the input of the system has greater influence on such information parameters, as double phase and square amplitude. These results suggest a promising use of the phase component of the signal as information parameter in the processing and simulation systems research of signals, reflected by extended surface.

References

[1] Helstrom C W 1968 *Statistical Theory of Signal Detection* (Oxford: Pergamon Press)
[2] Jackson L B 1989 *Digital Filters and Signal Processing* (Boston: Kluwer Academic Publishers)
[3] Antolovic D 2010 Radiolocation in Ubiquitous Wireless Communication (US: Springer)
[4] Oppenheim A V and Schafer R W2010 *Discrete-Time Signal Processing* (Harlow: Pearson)
[5] Bykov V V 1971 *Digital modeling in statistical radio engineering* (Moscow: Sovetskoe radio) (in Russian)
[6] Akimov P S *et al* 1984 *Theory of signal detection* (Moscow: Radio i svjaz') (in Russian)
[7] Timoshenkova Y S *et al* 2015 *Proc. CriMiCo* **1** 937 (in Russian)
[8] Hayes M H 1996 *Statistical Digital Signal Processing and Modeling* (New York: John Wiley & Sons)
[9] Karam L J *and* McClellan J H 1995 *IEEE Transactions on Circuits and Systems II: Analog and Digital Signal Processing* **42(3)** 4937837 DOI: 10.1109/82.372870
[10] Ustinova I *and* Kozlova A 2016 *IOP Conf. Ser.: Mater. Sci. Eng.* **142(1)** 012027 DOI:10.1088/1757-899X/142/1/012027

Assessment of the Charging Policy in Energy Efficiency of the Enterprise

E A Shutov[1], T E Turukina[2], and T S Anisimov[3]

[1]Associate Professor, National Research Tomsk Polytechnic University, Tomsk, Russia
[2]PhD Student, National Research Tomsk Polytechnic University, Tomsk, Russia
[3]Student, National Research Tomsk Polytechnic University, Tomsk, Russia

E-mail: shut63@mail.ru

Abstract. The forecasting problem for energy facilities with a power exceeding 670 kW is currently one of the main. In connection with rules of the retail electricity market such customers also pay for actual energy consumption deviations from plan value. In compliance with the hierarchical stages of the electricity market a guaranteeing supplier is to respect the interests of distribution and generation companies that require load leveling. The answer to this question for industrial enterprise is possible only within technological process through implementation of energy-efficient processing chains with the adaptive function and forecasting tool. In such a circumstance the primary objective of a forecasting is reduce the energy consumption costs by taking account of the energy cost correlation for 24 hours for forming of pumping unit work schedule. The pumping unit virtual model with the variable frequency drive is considered. The forecasting tool and the optimizer are integrated into typical control circuit. Economic assessment of the optimization method was estimated.

1. Introduction

The question of rational use of energy, as well as looking for ways to reduce the cost of acquisition is the most relevant for today. Steadily rising prices of energy sources [1] led to the fact that the energy part production expenses became disproportionately large part. As a result, a lot of industrial enterprises faced the challenge of energy- and resource intensity reducing of output product and services, i.e. the problem of energy saving. Energy resources consumption analysis at many enterprises shows that the problem solving has two directions: organizational-technical measures for eliminating the useless energy resource consumption, and the adoption of energy-efficient technologies and energy conservation equipment allowing to perform the same work content with less electrical energy consumption.

The choice of the most optimal tariff (price category) may act as organizational-technical measures for industrial enterprises. Within the Enactment of the Government of the Russian Federation 442, dated 04 May 2012, "On the retail electricity market performance, full and/or partial limitation of electrical energy use conditions" the customer depending on the usable maximum power at the point of balance sheet delineation is obliged to choose the most appropriate price category [1].

2. Formulation of the problem

The problem of choice is currently acute for the customers with the maximum power more than 670 kW. Since 01 July 2013 these customers can choose only PT with hourly energy accounting or price category with hourly energy accounting and planning, namely from third to sixth price category [1].

The price category should be chosen based on minimum electrical energy purchase costs that in turn depends on load curve and peak power. One of the most significant factor when choosing the PT is the willingness to planify of hourly energy for day-ahead market.

The lowest values of the unregulated price are formed for the fifth and sixth but when these price categories are chosen a customer is obliged to provide to the energy service company plan values of energy consumption for the next month with an hourly breakdown. The customer tariff includes the deviation cost of the planned energy consumption from the real value and the value of such deviations can be significant. For that reason for the water facility, which purchases electrical energy at the fifth and sixth price category, it is very important to have an accurate forecast of energy consumption.

About 90% of the electric power consumption by water facilities is required for pumping equipment work, speaking about Water Supply Company it is about 2 million kWh per month. In this regard, along with the economic dimension of such problem as decrease of cash expenditures, for Water Supply Company there is a perceived problem with the low energy efficiency of operation of pump units associated with the use of water flow control via valve throttling. It is known that replacing valve control with variable frequency drive control allows to obtain energy savings up to 60% in similar systems [2]. Such authors as Bose, K. Bimal, B.S. Leznov [3], I.A. Syromyatnikov, C. U.Brunner, V. B. Vozdvizhenskiy, Siskind, S. Charles, Spear Mike, Novak Peter dealt with the issues for energy consumption optimization by using of the variable frequency drive.

3. Experimental part

In practice the effectiveness of technological process is analyzed separately and the improvement of production effectiveness is considered separately. The goal of this work is joint considering of the energy efficiency indicators of technological process and the level of business processes. The use of classical algorithm of vector control of frequency drive may act as the instrument to achieve the result at the level of technological process. It is offered to use supervisory control and data acquisition (SCADA) system as the element of process visualization. This system will be available for the upper hierarchical structure of management. A classic variation of a frequency converter structure is the use of a vector or adaptive control the pump unit rate. In this way, both the adjustment control area and the controller's parameters or the controller structure variation are secured. It depends on the change of the process parameters or external disturbances acting on the control object.

Figure 1 shows the electricity load curve, which is characterized the pronounced nonlinearity, the presence of typical peaks and lows of the load.

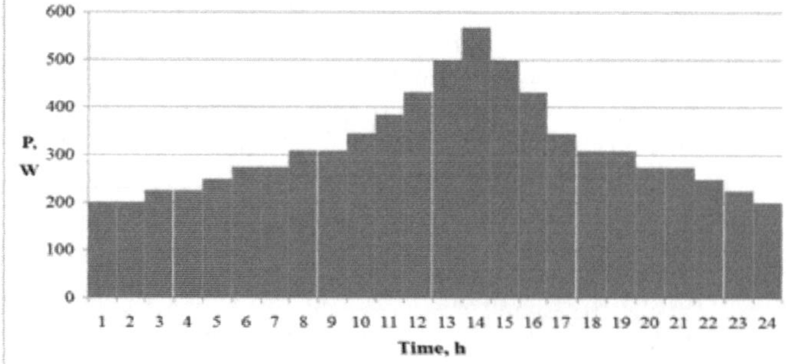

Figure 1. Power load diagram of the ventilation equipment.

In the experiment the study subject is the low pressure radial fan. The drive mechanism of the fan is the induction motor (IM) with a capacity of 0.55 [kW]. The installation also includes such units as

IOP Conf. Series: Materials Science and Engineering **189** (2017) 012026 doi:10.1088/1757-899X/189/1/012026

the frequency converter, the radial ventilator, the programmable logic controller, current and pressure sensors, an electrical power unit of the frequency converter, a bank of solid-state relay. The superordinate control level is a TRACE MODE SCADA system. Multiwindow interface of the system allowed to present mnemonic scheme of the technological process along with the monitor and control nodes, and form control procedure trends for the experimental unit. The mnemonic scheme is shown in Figure 2. It also shows the visualization example of the regulating characteristic taken from the experimental unit.

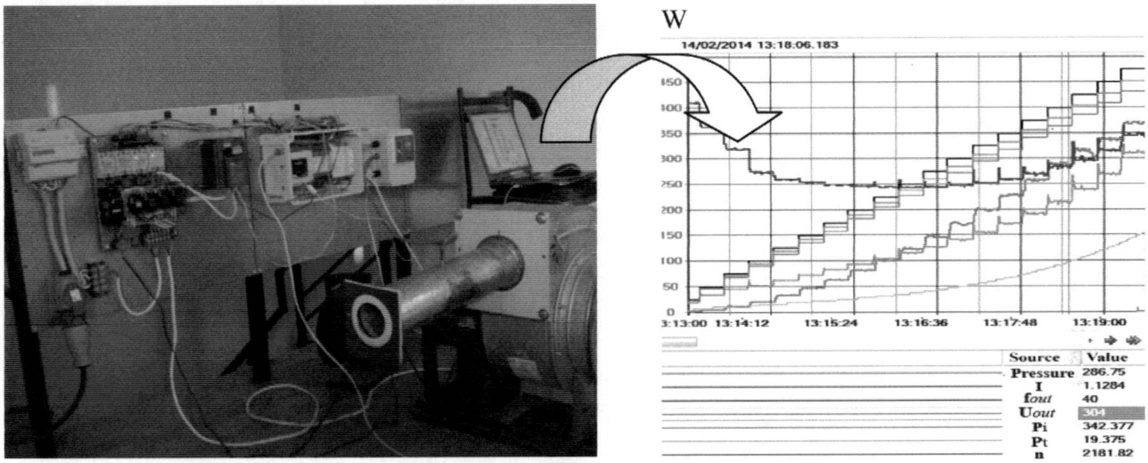

Figure 2. An experimental unit with characteristics of the fan operation.

Since output characteristics and the fan capacity are known, the experimental data sufficiency allow avoiding the use of analytic expressions, and the unit performance evaluation can be determined according to the database of the experiment. For the load curve (Figure 1) using pressure and power experimental loads, the pressure level in the system and energy consumption was determined for each day hour (Table 1).

Table 1. Calculated value for the load curve

Time	Power, [W]	Pressure, [Pa]	Energy consumption, [kWh]
00:00	200	142.31	200
01:00	200	142.31	200
02:00	224	171.35	224
03:00	224	171.35	224
04:00	248	198.31	248
05:00	274	228.33	274
		......	
23:00	200	142.31	200
Total sum			7601 [W]

The use of the variable frequency drive eliminates the inefficiency of the airflow control by means of the pneumo-latch throttling. At the implementation of the variable frequency drive, the pressure of 400 Pa is maintained in the system by means of rotation frequency varying of the driving unit. It leads to fan delivery change, that provides a constant pressure in the system when the air flow is changed. In percentage terms the efficiency of such approach is 79%.

IOP Conf. Series: Materials Science and Engineering **189** (2017) 012026 doi:10.1088/1757-899X/189/1/012026

4. Forecasting and optimization results

The introduction of the load forecasting tool and the optimization mechanism of the target function are therefore proposed. The goal of the optimization task is to reduce expenditures connected with electrical energy of the 3rd elevation pumping station taking into account the water intake curve, current tariff rates and process requirements (presence a reservoir) are made to the water pipe of the pumping and filtration station (PFS) – peak back–up boilerhouse (PBB).

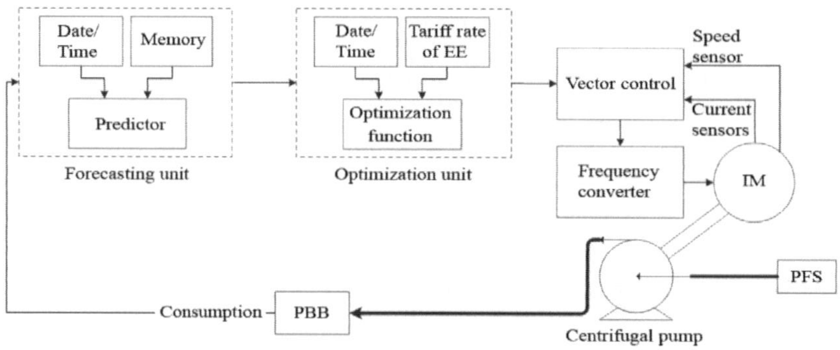

Figure 3. Flow-chart of the pumping station operation.

Organizational–technical measures leading to the elimination of the useless expenditure of power resources (the level of production control) are aimed at using the forecasting tool in the current performance of the enterprise. Such interpretation is conditioned by the electricity market. The forecast is a tool of the electricity market performance. The forecasting problem is highlighted, for example, in the works of the following authors: Nagasaka K, Mamun M. A., Hobbs N. J., Kim B. H., Lee K. Y., Mao H., Zeng X., Leng G., Zhai Y., Keane J.

The load curve of a water facility is characterized by stable cycles with the typical changes of energy consumption during the day (morning and evening peaks and the night minimum demand) and also during the week (the load decrease during the weekend due to the absence of water consumption).

The first step of any forecasting is the source data analysis. The signal noiseness and random components, that have no value for formation forecast model, are excluded by special prepare the source data. It can be wavelet-analysis method, which is currently one of the most advanced data analysis technology. Figure 4 shows the result of the wavelet decomposition. The fundamentals and the possibility of use wavelets for the times series forecasting are given in the papers of the following authors: Yao S.J., Song Y.H., Zhang L.Z., Cheng X.Y., Smolentsev N.K, Yakovlev A.N., K. Blatter, Daubechies, Mallat S., Vorobyev V.I., Gribunin V.G.

The results of Ljung-Box Q-test indicate that the source data is not random variables and the energy consumption forecasting is possible. The method of neural networks with the exposure into a neural network logical basis was chosen as the forecasting method [4–8]. The time frame was provisionally divided into 3 equal sequences to make independent forecasting. Thus, the range of valid values is formed, where the forecast value is determined by laws of the fuzzy logic [9].

The network structure is the three-layer neural network with the following parameters: $v=3$ is each value of the fuzzy variables, $n=v\cdot w=12$ is the number of the first layer neurons, $m=9$ is the number of the second layer neurons, $k=v=3$ is the number of the third layer neurons, 67 is the number of stages for the model adaptation [10].

Figure 5 shows the result of forecasting with the precision up to 0.55%.

When the enterprise operating at the sixth price category, the obtained accuracy of forecast save 1426 rubles a month, which is 0.1% of the total value of the payment.

The most efficient instrument to reduce the energy charge for an enterprise is a joint accounting of energy efficiency of both the technological process level and the production control level.

Water supply organizations are characterized by a steady load profile of water draw-off which is well correlated with the energy consumption curve. The reduction of correlation, including the reduction of the power demand during peak hours in the region is achieved by using the forecasting tool. The energy consumption reduction is efficient, when the medium-term forecasting characteristic of water draw-off is known and the electric drive of the pump operates without taking account the received perturbation influence by the sensor of the water supply system and chooses the optimal control trajectory of the reservoir filling point.

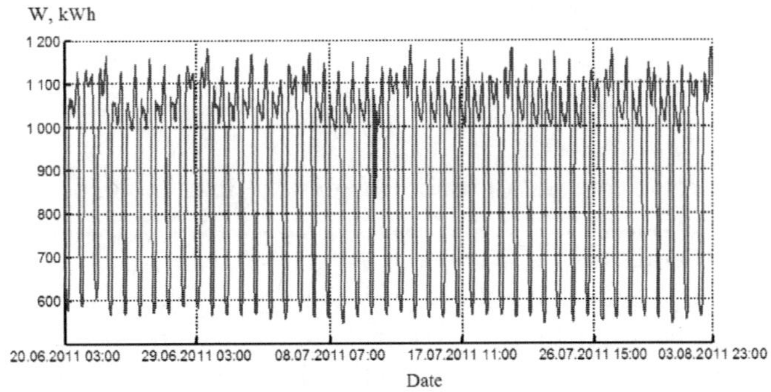

Figure 4. The reconstructed time frame of the load curve of the 3rd elevation pumping station

Operational changes of the pumping unit occur when the forecasting tool and optimizer are implemented in a typical control circuit with the variable frequency drive. But it depends on the cost of energy for every hour and under condition of keeping the customer load curve (in this case, peak standby boiler plant). It can be achieved through using a service water reservoir at the peak standby boiler plant, which storages water at minimum energy cost hours and yields it on the needs of a customer at maximum water draw-off hours.

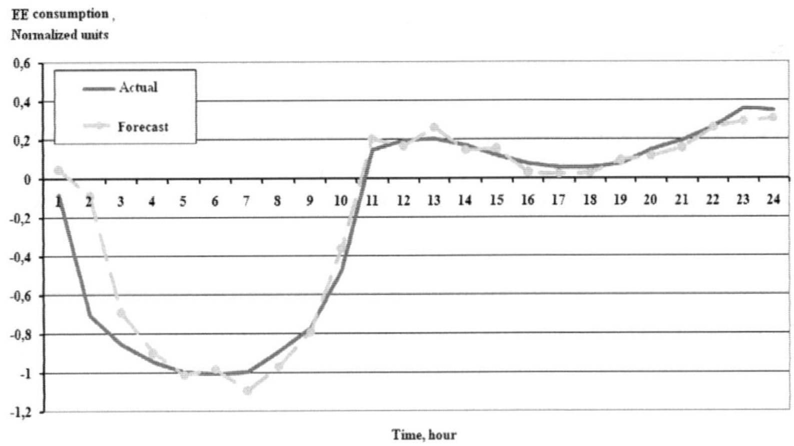

Figure 5. The result of the energy consumption forecasting by a method of the neural network.

As the result, the purchase amount of energy and power at maximum price decreases and the share of purchase at minimum price increases. The load change takes place by optimizer at maximum energy cost hours and peak value of capacity (08:00).

Optimization results depend on the control method of the pumping unit are shown in the Table 2. By the results of research, the conclusion that forecasting optimization of the 3rd elevation pumping station performance has resulted in the reduction of power costs, but at the same time power requirement is not decreased (406.37 kWh), but also decreased (411.2 kWh). There is no efficiency on the technological process level, however, at the production level the efficiency of this regulation (the reduction of the energy purchase costs) is 26%.

Table 2. Electrical energy cost and consumption

Control method	Energy consumption per calendar day, [kWh]	Power costs per day, [rubles]
Varying the speed of a pump	406.37	1180.70
Varying the speed of a pump based on optimization function and consumption forecast	411.20	870.58

5. Conclusion

The results obtained attest to the forecast-optimization control in combination with the pricing policy allow to reach significant fund savings up to 60% in terms of peak power values and up to 12% in terms of the energy cost. Based on the experimental data it can be concluded, that the methods engineering of forecast-optimization control of water-supply engineering tract equipment including level-to-level relationship of control eliminates correlation between the load curve of the peak standby boiler plant and 3rd elevation pumping station and the methods engineering sets new relations between the change of the tariff rate and the pumping unit operating mode. The scope of application of forecast-optimization control is enterprises with a nonuniform load profile and a possibility of storing the product (water, gas) [11]. The time differentiation of energy rates is the necessary condition.

References

[1] *About Functioning of the Retail Markets of Electric Energy, Full and (or) Partial Restriction of the Mode of Consumption of Electric Energy* (approved by the Government of the Russian Federation, an edition from 26/12/2016 No. 442) Retrieved from http://www.consultant.ru/document/cons_doc_LAW_130498/ (in Russian)

[2] Bethoux O *et al* 2017 *Electric Power Systems Research* **142** 176–189 10.1016/j.epsr.2016.09.014

[3] Leznov B S 2006 *Energy Saving and Controlled Drive in Pump Sand Blowing Machinery* (Moscow: Energoatomizdat) (in Russian)

[4] Narayan S *and* Smyth R 2015 The financial econometrics of price discovery and predictability// *International Review of Financial Analysis* **42** Pages 380–393 DOI: 10.1016/j.irfa.2015.09.003

[5] Lin C T *and* Lee C S G 1994 *Supervised and unsupervised learning with fuzzy similarity for neural-network-based fuzzy logic control systems* (New York: Van Nostrand Reinhold)

[6] Homik K *et al* 1989 *Neural Networks* **2(5)** 359–366

[7] Zadeh L A 1965 *Information and Control* **8(3)** 338–353 DOI: 10.1016/S0019-9958(65)90241-X

[8] Costa T V *et al* 2009 *Computer Aided Chemical Engineering* **27** 1407–1412 DOI: 10.1016/S1570-7946(09)70625-X

[9] Giuliani M *et al* 2016 *Environ Model Softw* **85** 293–298 DOI: 10.1016/j.envsoft.2016.08.015

[10] Babinovich D E *and* Shutov E A 2014 *WIT Transactions on Ecology and the Environment* **190** 501–510 DOI: 10.2495/EQ140481

[11] Sumarokova L *et al* 2016 *MATEC Web of Conferences* **92** 01025 DOI: 10.1051/matecconf/20179201025

Deformable Polymer Dielectric Films in Phase Light Modulators

P A Luchnikov[1], V V Vetrova[1], N P Zubkov[1], O A Surzhikova[1], and N I Chernova[2]

[1]Senior Researcher, Moscow Technological University (MIREA), Moscow, Russia
[2]Professor, Moscow Technological University (MIREA), Moscow, Russia

E-mail: xamdex@gmail.com

Abstract. Experimental study of the deformation amplitude of the dielectric polymer gel films directly right in the display device of optical information was conducted. The deformation amplitude of the gel film was measured during its change by controlled constant voltage on the electrodes. On the basis of polymer gel deformational layer the display devices of optical information and registration of electrostatic potential of charged dielectric thin films in the presence of defects are considered.

1. Introduction

The devices based on polymer gel atmospheres (GA) are developed to display of information in optical form [1, 2]. The devices based on GA are developed in two directions of recording information: by electron beam and at the electrode controlling. At the electrode controlling the service life of polymeric GA is increased; the size and power of the device are reduced. It is also important to know the necessary values of GA deformation for getting the maximum light output.

2. Display device of optical information record

Figure 1a shows scheme of the display devices of optical information on basis on polymer deformational gel layer.

The device works in the following way. Capacitor 1 displays filament body of the light source 2 on the opening of input diaphragm 3; after that through the lens 4 a transparent conductive layer 6 and parallel light stream, directed on the GA 5 through the prism of total internal display 7, are formed. Next, the light, reflected from the interface of "GA – air gap", secondarily passes through GA, the electrically conductive layer 6 and the prism of total internal display 7, and then focuses on the opaque output diaphragm 12 by the lens 8. Control voltages from source of control signals 10 are fed on the strip electrodes 9 of the raster on the second glass plate 11. GA surface is deformed in accordance with the voltage on the electrodes. GA relief changes the direction of light propagation; rejected light beams will pass to the screen, bypassing the opaque output diaphragm 12. Lens 8 focuses this light on the screen 13 in the form of light strips. The brightness of the light strips is proportional to the amplitude of the control voltages. The light output of the device was measured as the dependence of the light line brightness on the screen on the voltage value on the control electrodes.

Two semitranslucent 14, 15 and reflecting mirrors 16 are included in the scheme for measuring the amplitude of the deformations of GA surface. These mirrors create two arms of the Mach–Zehnder interferometer; they form an interference picture on the screen 13 during setting. Since the deformable gel layer is in one arm of the interferometer under the action of the control voltage, the interference picture shows the increasing phase incursion of light due to change of the deformation amplitude. A source of coherent radiation (helium-neon laser) was used as radiation source.

Figure 1b shows dependence of the light output (ρ) on the constant electrical voltage U on the electrodes. It follows from the graph that the sensitivity of the method of defocusing is considerably higher sensitivity of the method of dark field on 100 V.

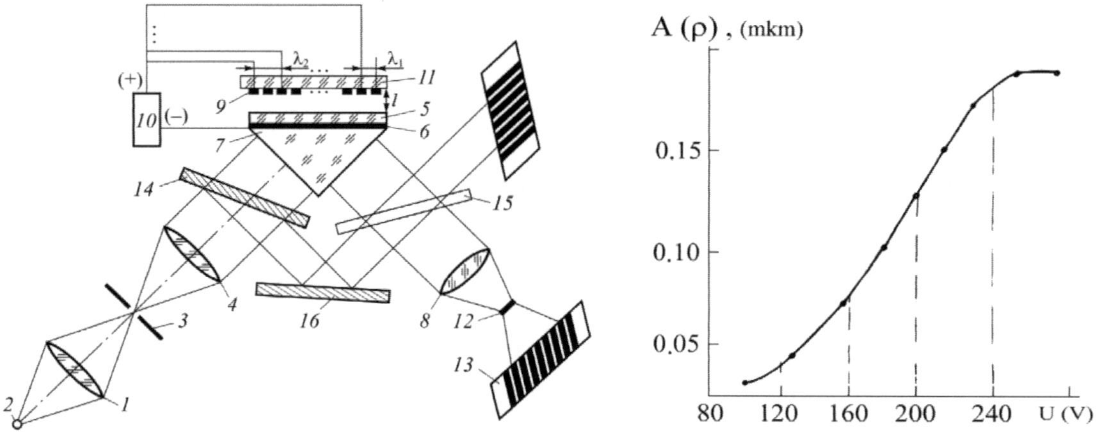

a *b*

Figure 1. Scheme of the display devices of optical information on basis on polymer deformational gel layer (*a*) and dependence of the light output (ρ) on the constant electrical voltage U on the electrodes (*b*).

Control voltage amplitude U increases sharply with increasing of air gap between the GA and strip electrodes 9 (Figure 1a); it is a condition for getting of maximum light output. Light output function equal to 0 at the gap ~ 100 μm at any voltages between control and rasterized electrodes. Reduction of light output is possible to 0 at simultaneous feeding a control voltage on the electrodes of raster 9, when the voltage has a value, close to the breakdown voltage of the working air gap between the electrically conductive substrate 6 GA film 5 and the electrodes 9 (with the gap of 40 μm). This is because the density of the ponderomotive forces of harmonic components decreases exponentially with the increasing of the air gap; the light output depends on the density of the ponderomotive forces of harmonic components. At the same time, the density of the ponderomotive forces of constant component varies slightly with the increasing of the gap. Light output decreases sharply for a time of 10÷20 seconds at feeding the control voltage on all the control electrodes and electrodes of raster 9. This change of light output occurs because of the leakage of the charge in the electrode gaps due to the final value of the electrical conductivity of the substrate and potential leveling of control plane of the film, respectively. If raster electrodes are connected to zero potential, then slow decreasing of light output occurs exponentially at disconnecting of the source of control voltage from the control electrodes. This phenomenon can be used as an electrostatic memory.

Light output and GA strain amplitude can be measured for the same conditions. The light output curve can be represented as the dependence of strain value on the control voltage; it shows that a small increment of amplitude of the order of 0.18 μm is required for getting maximum light output. This is consistent with theoretical estimates of phase increment taking into account the double passage of light through the surface relief of GA [2]. Relative elongations do not exceed 1% of the polymeric composition of GA with thick of 20 μm. Therefore, the equations of continuum mechanics in the

IOP Conf. Series: Materials Science and Engineering **189** (2017) 012027 doi:10.1088/1757-899X/189/1/012027

linear approximation can be used at the calculation of deformation amplitude value. Performed studies show a good optical quality of the spatial-time light modulators based on polymer GA.

3. Control device of defects in thin films

The need to control the heterogeneity of the surface properties and defects of technological thin layers of dielectrics appears during development and study of films in microelectronics [3, 4], electret [5–7] and the active thin film materials [8–12]. One of effective methods of testing is a principle of spreading of excess charge on a controlled surface; in this case the charge drains arise in the locations of defects (or inhomogeneities). It leads to considerable heterogeneity of potential relief of surface film of charged dielectric. Heterogeneity of potential relief of surface indicates location and defect size.

Vibrating electrode method with compensation of external electric field is used for registration the potential relief of surface of charged dielectric materials, for example, electrets [13]. The method has high measurement accuracy with an error not exceeding ±1%. In this case, vibrating electrode (as a probe) is necessary to implement structurally with small area that is making significant edge effect of the electrode in the measuring cell; it leads to significant increasing measurement error at low resolution.

The above-described electro-optical principle of gel dielectric layer of GA (as a probe) in measuring cell can be used to solve the problem of registration of potential relief with high resolution. Layer deformation of GA is carried out by external electric field E_0 as charged dielectric in system o flat capacitor and by supplying external voltage into electrodes of capacitor. It allows studying the picture of charge distribution on the surface in the form of light inhomogeneities with subsequent signal processing.

Figure 2 shows schematic diagram of measurement system based on electro-optical sensor (EOS) with GA, where 1 is charged dielectric layer of sample, coated on hypotenuse side of prism, 2 is sensitive GA, 3 is transparent electrically conductive electrode, 4 is photo receiving optical device of recording by dark field method, 5 is prism, 6 is source of the plane-parallel light beam, 7 is supply source, 8 is electrode, 9 is regulated constant-voltage source, 10 is defect in the film, 11 is display device of information.

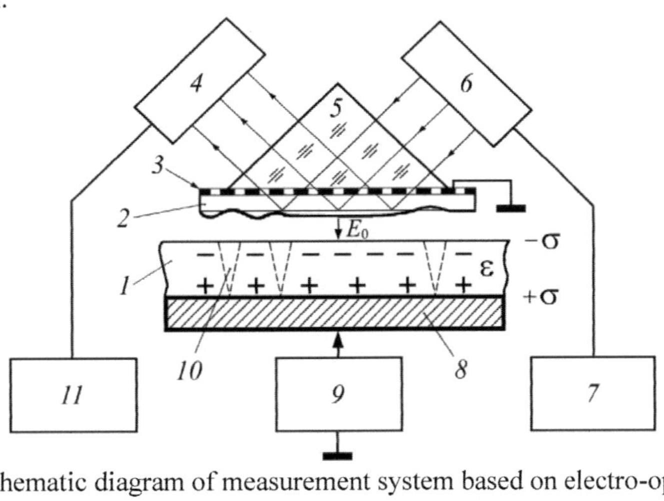

Figure 2. Schematic diagram of measurement system based on electro-optical sensor.

The device (Figure 2) works as follows. Charged sample of film dielectric 1 with thickness L on the substrate 8 is placed in cell of flat capacitor formed by plane-parallel electrodes 3 and 8 with fixed gap d_0. Electric field E_0 is generated in gap d_0 of the capacitor by dielectric charge with effective density of surface charge σ. It causes deformation of the surface of GA 2 by ponderomotive forces of interaction. Plane-parallel light beam from source 6 passes through prism 5, optically transparent electrode 3 and GA 2; it is reflected from GA boundaries, returns to the prism 5 and then falls into photodetector 4 (PD), for example, CCD camera.

Transformation of distribution picture of electric potential of charged surface of studied film 1 occurs in EOS with gel layer of GA (electric signal $\sigma \rightarrow U_0 \rightarrow E_0$ into output optical signal in form of light image). PD 4 registers and visualizes a picture of light field on the screen of display device of optical information 11.

Surface charge flows on the substrate 8 is significantly faster in the defect region 10 of film 1 and forms a region of reduced surface charge density σ. Gel layer 2 of EOS of electric field has a maximum value of deformation in non-uniform charge region on the dielectric surface 1, i.e. in film defect region. Therefore, PD 4 registers more bright fluorescence in this region of picture of light display.

Thus, charge regions on surface with greater or lesser charge density appear on picture of light display as different intensity of emission in this defect region. Brightness in predetermined region of picture can change to certain light intensity by varying the constant electric voltage from source 9 on the electrode 8. Voltage value U_K of compensation field E_0 is recorded at full compensation of the electric field in gap at E_0=0; charge density σ in the controlled region is calculated according to equation $\sigma = \varepsilon\varepsilon_0 \cdot U_K / L$. The light picture of field shows complete picture of potential relief of controlled dielectric surface in the form of two-dimensional light field with presence of light spots corresponding to defects 10 of investigated layer 1 of film dielectric. Use of standard optic in control system (with the gap between sample and GA) allows to determine defects in the form of dots and spots of size less than 50 microns at resolving of potential of dielectric surface of not more ±5 V.

5. Conclusion

Experimental studies of deformation amplitudes of gel films directly in optical pickup device of potential relief of charged dielectric show that it is possible to control parameters of sensitivity and registration accuracy of optical information by feeding external constant electric voltage.

Device of registration and defects control in thin and nano-sized dielectric films in microelectronics is proposed on the basis on the studied phenomenon of gel layer deformation under action of the ponderomotive forces of electric field.

Discussed method also allows carrying out research of potential relief picture of charged dielectrics.

Acknowledgments

This work was supported by the Ministry of Education and Science of the Russian Federation (No 3.8074.2017/BCh, basic part of the State Task).

References

[1] Gurevich S B (Ed) 1977 *Spatial light modulators* (Leningrad: Nauka) (in Russian)
[2] Guscho Yu P 1992 *Physics of reliefographic* (Moscow: Nauka) (in Russian)
[3] Aleutdinov A D *et al* 2015 *IOP Conf. Ser.: Mater. Sci. Eng.* **81(1)** 012069 DOI: 10.1088/1757-899X/81/1/012069
[4] Yakimov E *et al* 2017 *J. Phys.: Conf. Ser.* **783** 012061 DOI: 10.1088/1742-6596/783/1/012061
[5] Luchnikov P A *et al* 2016 *IOP Conf. Ser.: Mater. Sci. Eng.* **110(1)** 012081 DOI: 10.1088/1757-899X/110/1/012081
[6] Pevtsov E Ph *et al* 2017 *IOP Conf. Ser.: Mater. Sci. Eng.* **168(1)** 012095 DOI:10.1088/1757-899X/168/1/012095
[7] Luchnikov P A 2015 *Russ Phys J* **57(9)** 1239–1245 DOI: 10.1007/s11182-015-0369-0
[8] Ragachev A A *et al* 2015 *Appl. Surf. Sci.* **351(1)** 811–818 DOI: 10.1016/j.apsusc.2015.06.008
[9] Lysenko E N *et al* 2014 *J Therm Anal Calorim* **115** 1447–1452 DOI: 10.1007/s10973-013-3456-x
[10] Luchnikov P A *et al* 2017 *IOP Conf. Ser.: Mater. Sci. Eng.* **168(1)** 012092 DOI:10.1088/1757-899X/168/1/012092

[11] Piliptsou D G *et al* 2017 *IOP Conf. Ser.: Mater. Sci. Eng.* **168(1)** 012103 DOI:10.1088/1757-899X/168/1/012103
[12] Surzhikova O A 2015 *IOP Conf. Ser.: Mater. Sci. Eng.* **81(1)** 012098 DOI: 10.1088/1757-899X/81/1/012098
[13] Sessler G M (Ed) 1980 *Electrets* **33** (Berlin: Springer Verlag, New York: Heidelberg)

IOP Conf. Series: Materials Science and Engineering **189** (2017) 012028 doi:10.1088/1757-899X/189/1/012028

Method of Calculating the Correction Factors for Cable Dimensioning in Smart Grids

M Simutkin[1], V Tuzikova[2], J Tlusty[3], V Tulsky[4], and Z Muller[5]

[1]Ph.D Student, Moscow Power Engineering University, Moscow, Russia
[2]Ph.D Student, Czech Technical University, Prague, Czech Republic
[3]Deputy Head of the Department, Czech Technical University, Prague, Czech Republic
[4]Head of the Department, Moscow Power Engineering University, Moscow, Russia
[5]Head of the Department, Czech Technical University, Prague, Czech Republic

E-mail: tuzikval@fel.cvut.com

Abstract. One of the main causes of overloading electrical equipment by currents of higher harmonics is the great increasing of a number of non-linear electricity power consumers. Non-sinusoidal voltages and currents affect the operation of electrical equipment, reducing its lifetime, increases the voltage and power losses in the network, reducing its capacity. There are standards that respects emissions amount of higher harmonics current that cannot provide interference limit for a safe level in power grid. The article presents a method for determining a correction factor to the long-term allowable current of the cable, which allows for this influence. Using mathematical models in the software Elcut, it was described thermal processes in the cable in case the flow of non-sinusoidal current. Developed in the article theoretical principles, methods, mathematical models allow us to calculate the correction factor to account for the effect of higher harmonics in the current spectrum for network equipment in any type of non-linear load.

1. Introduction

There are currents in neutral conductors for feeding non-linear consumers. The appearance of these currents due to the lack of phase shift at higher harmonics are multiples of the three and fundamental frequency. In some cases, the RMS current in the neutral conductor may reach 1.5-2 times higher than in the phase conductor [1]. Overloading of neutral cable conductor caused by higher harmonic currents leads to a further increase of temperature above the limit of provided by manufacturer, which accelerates the insulation aging of the cable and therefore reduces its lifetime [2, 3]. Current harmonics cause additional heating of the equipment; this should be taken into account for selection the conductor cross-section size and control allowable currents during operation [4-6].

2. Current Harmonic Spectra of Individual Power Consumers and Their Aggregation

To determine the possible effect of higher harmonics on cable lines it is necessary to assess the levels of interference caused by nonlinear electro receivers individually and collectively. For this purpose, there were performed experiments under certain types of power appliances with nonlinear voltage-current characteristics, which are most commonly used in residential and administrative buildings. The results show that these types of power appliances create high levels of distortion in the current waveform (third harmonic level reaches 80...90 [%], the fifth reaches 45 [%]), while all the above power-consuming equipment comply with the standard for emissions higher harmonic components of electro receivers with consumption current less than 16 [A] [7], and therefore can be freely used by consumers in the country. Thus, the requirements of the standard [7] can't provide such levels of higher harmonics, which do not affect the electricity network. Measurements were performed on real objects (shopping center, specializing in the sale of computer equipment, office building and houses) to determine the possible values of current harmonic and spectra. The results are shown in Table 1. These spectra were used to calculate values for the correction factors for specific types of nonlinear consumers.

Table 1. Range of harmonics current to different types of consumers to the maximum load.

Number of harmonic	Shopping center $I_{(n)}$, [%]	Administrative building $I_{(n)}$, [%]	Residential building $I_{(n)}$, [%]
1	100.0	100.0	100.0
3	31.2	29.3	10.3
5	18.5	13.9	6.3
7	12.7	10.2	4.8
9	8.6	15.2	5.5
11	5.4	9.5	4.6
13	3.2	8.2	3.2
15	2.7	3.9	1.9
17	1.7	4.6	-
19	1.1	2.2	-
21	0.7	1.4	-

The flowing of non-sinusoidal current through the phase conductor of cable generates heat; power is determined by the loss of active power at the fundamental frequency and high harmonics [8]:

$$P_{\Sigma}^{ph} = I_1^2 \cdot R_1 + \sum_{n=2}^{40} I_n^2 \cdot R_n \quad (1)$$

where I_1 and In are currents of the fundamental frequency and high harmonics, R1 and Rn - resistance at the fundamental frequency and high harmonics.

There are flowing harmonic currents of zero sequence in the neutral conductor under condition of balanced nonlinear load. This releases heat is power source, which is determined by the following formula:

$$P_{\Sigma}^{ph} = 3\sum_{n=3,9,15} I_n^2 \cdot R_n. \quad (2)$$

Active resistance of the conductor on the n-th harmonic (Rn) in (1) and (2) for n ≥ 3 is defined by the formula [9]:

$$R_n = R_1 \cdot (0.187 + 0.532 \cdot \sqrt{n}), \quad (3)$$

where R1 is the resistance of the conductor at the fundamental frequency currents, n is number of the harmonic. The generated heat inside the cable, which is laid in the ground, is transmitted by the phenomenon of thermal conductivity into the surrounding space. The steady heat cable mode is described by Kirchhoff Law equation:

$$\alpha \cdot \nabla^2 T = 0, \quad (4)$$

where ∇ is the Laplace operator. To develop a mathematical model for the study of the cable and thermal processes in it under the influence of harmonic currents that determine the capacity of cable lines, the most common types of cables were chosen and for each of them was composed mathematical model in the software Elcut [10]. Using this program we calculate the temperature field of the cable according to the equation (4) for the stationary mode using finite element method.

4. Method for Determination of Ampacity Reducing due to Harmonics

General active power losses in the cable are the sum of tripled active power losses in the phase conductor and losses in the neutral conductor. Using the algebraic manipulation and formulas (1) - (3), we get the formula of general active power losses in the cable into a form which differs from the loss at the fundamental frequency, only the additional loss factor (K_{add})

$$P_{\Sigma} = 3 \cdot I_1^2 \cdot R_1 \cdot K_{add} \tag{5}$$

$$K_{\text{add}} = 1 + \sum_{n=2}^{40}(K_{I_n})^2 \cdot A_n + 3 \cdot \frac{R_1^n}{R_1} \sum_{\substack{n=6k-3 \\ n=3,9,15}}^{40}(K_{I_n})^2 \cdot A_n, \tag{6}$$

where K_{I_n} the current value of n-th harmonic component as a fraction of the current fundamental frequency, $A_{\text{n}} = 0{,}187 + 0{,}532 \cdot \sqrt{n}$.

Then, we introduce the concept of equivalent current (I_{eq}). Equivalent current is the current of the fundamental frequency which flows through the three conductors and it generates the same amount of heat as through a non-sinusoidal current flowing in the three conductors and the neutral conductor. By their energy essence, the notion of an equivalent current is the transition from four sources of heat (three cores and neutral) to the three sources (three conductors), and it is assumed that all heat from power distorted currents is replaced by an equal value of active power losses created by the current at fundamental frequency, as in terms of energy no matter which of the current heats the conductor: 50 Hz sinusoidal or non-sinusoidal. Given entered term formula (5) can be written as

$$P_{\Sigma} = 3 \cdot I_1^2 \cdot R_1 \cdot K_{add} = 3 \cdot I_{eq}^2 \cdot R_1 \tag{7}$$

where $I_{\text{eq}} = I_1 \cdot \sqrt{K_{add}}$.

As a result, the desired correction factor to the permissible continuous current of the fundamental frequency (K_{hh} is factor of higher harmonics), which recognizes the influence of harmonic currents flowing through the cable is

$$K_{\text{hh}} = 1/\sqrt{K_{add}} \tag{8}$$

This ratio is intended for selecting the power cable, as a correction value table of permissible continuous currents to prevent overheating of the cable insulation. For the RMS value of the current it is better to use other coefficient $K_{\text{n-1}}$, which value is determined by the formula

$$K_{n-1} = K_{hh} \cdot \sqrt{1 + \sum_{n=3}(K_{I_n})^2} \tag{9}$$

4. Example of Correction Coefficients Calculation

Power cable ASB 4x150 feeds nonlinear load (shopping center). It is necessary to determine the value of the allowable long-term current at the fundamental frequency, RMS value currents in phase and neutral conductor and the coefficient of higher harmonics in which the cable insulation does not overheat. Phase current harmonic spectrum is shown in Table 1. Resistance of cable conductor $R_1 = R_1^n = 0.243\Omega/km$ Manufacturer value of permissible continuous current is I_{perm}^{old}= 281 A [11]. From the known spectrum of harmonics initially determined by the necessary formula (6), components: A_n,K_{In}^2, as well as their sum of the harmonics of zero sequence occurring at neutral, across the spectrum of the current phase in the conductors. These amounts equal to $\sum_{n=2}^{40}(K_{I_n})^2 \cdot A_n = 0.2045$; $\sum_{\substack{n=6k-3 \\ n=3,9,15...}}^{40} (K_{I_n})^2 \cdot A_n$=0.1229. Further the calculated K_{add} coefficient, using formula (6), is $K_{add} = 1 + 0.2045 + 3\frac{0.243}{0.243} \cdot 0.1229 = 1.573$. With the known level of magnification of additional losses K_{add}, determination of the correction factor K_{hh} does not make a lot of effort

IOP Conf. Series: Materials Science and Engineering **189** (2017) 012028 doi:10.1088/1757-899X/189/1/012028

$$K_{hh} = \frac{1}{\sqrt{K_{add}}} = \frac{1}{\sqrt{1{,}573}} = 0.797.$$

Permissible continuous current is determined by the formula (9)

$$K_{n-1} = 0{,}797 \cdot \sqrt{1 + 0.10552} = 0{,}838.$$

Allowable current value of long-term fundamental frequency corrected for harmonic currents power $I_{perm}^{new} = I_{perm}^{old} \cdot K_{hh} = 281 \cdot 0{,}797 = 224.04$ [A]. Table 2 shows results of calculation of coefficients for oferent cable design, supplying shopping center.

Table 2. Results of calculation of coefficients for oferent cable design, supplying shopping center.

Cable brand	ASB 3x120	ASB 3x120+1x70	ASB 4x120	APvbbShp 4x120
K_{n-l}	0.695	0.771	0.875	0.875

On the picture of the thermal field (Figure 1) it is shown that the maximum temperature of the hottest point of the insulation is 78. 5 °C. Thus, using the method described above can be relatively easy to obtain the correction values of the coefficients. However, this method introduces an error in determining the value of the correction coefficient. This is due to the fact that heat sources are separated by an insulation which is a thermal resistance. In fact, if the current RMS value in neutral conductor is commensurate to the phase current and a neutral conductor cross-section is less than the cross-section of the phase conductor, the heat power in neutral conductor exceeds the value of the phase heat power. As a result, the most heated point of core insulation will be near neutral conductor.

5. Use of the Described Method with Respect to the Measured Spectra of Harmonics

Relative to previously measured spectra of harmonics for three types of consumers (Table 1) were obtained the following correction factors to the current values of the long-term allowable currents of the fundamental frequency (Table 3).

Table 3. Values of the correction factor for power cables 380 [V], supply different types of non-linear load.

Type of load	Value of factor K_{n-l}	
	For cables with phase conductor cross section greater than neutral	For cables with the same cross section neutral and phase wires
Shopping center	0.77	0.88
Administrative building	0.76	0.88
Residential building	0.96	0.97

After the calculations and the subsequent analysis of the results were made the following conclusions:

1) The value of the correction factor for the four-wire cable with smaller neutral cross-section size is less than for the cables with the same cross-section sizes of neutral and phases conductors. The wider range of harmonics generated by the nonlinear load, and the more the RMS value of current harmonics (especially harmonic multiples of three), the greater the difference of the above corrections for cable cross-sections for the same phase conductors.

2) Regardless of the phase conductor cross-section size, correction factor for all sections will have the same value.

3) An office building and shopping center have about the same impact on the cable in distribution

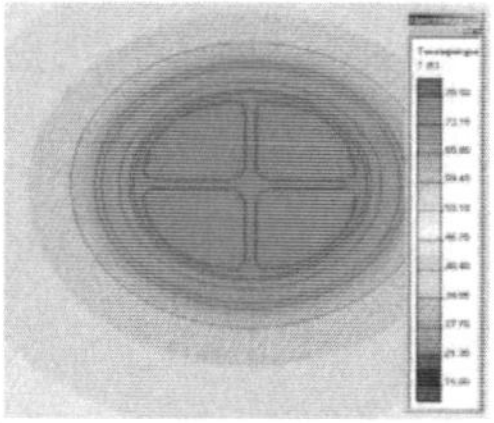

Figure 1. Picture of the thermal field in the cable ASB 4x150.

network 380 V due to the proximity of non-linear power appliances that are located at the consumer.

Also it has been demonstrated using calculations if the any changes for the value of the coefficients for the different types of cable designs claimed for one cross-section. Calculation with similar ratios for other cable types is considered at the example of the cable with 120 mm2 cross section size of the phase conductor. For comparison, previously selected three types of cables used for laying in trenches:

- ASB 4x120 cable with paper-oil insulation;
- ASB 3x120 +1x70 cable with paper-oil insulation;
- ASB 3x120 cable with paper-oil insulated neutral conductor, performed on the cable armature;
- APvBbShp 4x120 cable with XLPE insulation.

The results of calculation and simulation (Table 3) showed that the value of the correction coefficient is not affected by the type of insulation, but defined by cross-section only by the neutral conductor.

Developed method may calculate only coefficients for the ampacity reduction due to harmonic currents, but as the heat problem is not solved it is impossible to know the temperature of the cable. Thermal resistance between the conductor and the sheath (S1) for three core cables with sector conductors and core wrapping insulation is determined by the formula

$$S_1 = \frac{\rho_T}{2\pi} \cdot 3 \cdot \left(1 + \frac{3 \cdot t}{2\pi \cdot (d_x + t) - t}\right) \cdot \ln\left(\frac{d_a}{2 \cdot r_1}\right); \quad (11)$$

where ρT is the thermal resistance of insulation, [K · m / W]; d_a is core diameter, [mm]; r_1 is the radius of the circle circumscribed around the cores, [mm]; *dX* is the diameter of a circular conductor with a cross sectional area and the degree of compaction, that the shaped conductor [mm]; *t* is thickness of insulation between core and metal wrapping, mm.

Thermal resistance between the sheath and armor

$$S_2 = \frac{\rho_T}{2\pi} \cdot ln\left(1 + \frac{2 \cdot t_2}{D_s}\right); \quad (12)$$

where t_2 is the thickness of the cushion under armor, [mm]; DS is outer diameter of surface, [mm]. Thermal resistance of the outer protective coating

$$S_3 = \frac{\rho_T}{2\pi} \cdot ln\left(1 + \frac{2 \cdot t_3}{D_a}\right); \quad (13)$$

where t_3 is the thickness of the protective coating, [mm]; D_a is outer diameter of armor (for unarmored take outer diameter element, usually located directly under the armor, i.e. shell, screen or pillows), [mm].

Thermal resistance of the environment for insulated cable laid in the ground is

$$S_4 = \frac{\rho_T}{2\pi} \cdot ln\left(\frac{2 \cdot L}{D_e} + \sqrt{\left(\frac{2 \cdot L}{D_e}\right)^2 - 1}\right); \quad (14)$$

where *L* is distance from the surface of the ground to the cable, [mm]; D_e is cable outer diameter, [mm].

Temperature of core insulation for four-wire cable will be equal

$$\tau_c = P_c' \cdot S_1 + 3 \cdot P_c'(S_2 + S_3 + S_4) + \tau_a, \quad (15)$$

where τa is ambient temperature, [K].

For three-core cable with neutral conductor formed on the sheath

$$\tau_c = P_c \cdot S_1 + 3 \cdot (P_c + P_n)(S_2 + S_3 + S_4) + \tau_a \quad . \quad (16)$$

Formulas (15) and (16) show that the temperature of core insulation is directly dependent on the heat with the constant component that is equal to the ambient temperature. Using these expressions does not introduce a large error in the final result, which was confirmed by the results of calculation and subsequent comparison with the results of mathematical simulation. Error was less than 2%. This is primarily due to the fact that the greatest resistance is the thermal resistance of the earth; it is about 60% of the total. Thus, the heat equivalent conversion from the conductors cannot make a significant error.

6. Summary

The proposed method of calculating the correction factors can be applied on the stage of cable selection which supplies residential areas, and during periodic monitoring of the current in distribution networks. The value of the correction factor depends on cross-section size of neutral and harmonic current spectrum. Regardless of the phase conductor cross-section size, correction factor for all cross-sections will have the same value. The approach can be used for three and four core cable lines for all types of insulation voltage up to 20 kV. Using of developed method in special software will protect distribution networks and gives opportunity for management electricity consumption and operating network ampacity.

Acknowledgments
The research is supported by the Russian Ministry of Education and Science (project 2.541.2014K).

References

[1] Szydlovsky A *and* Zarkin A 2005 *Higher harmonics in low voltage networks. edition* (Kiev: Naukova Dumka) ISBN 966-00-0471-0
[2] Shafiullah G M *et al* 2013 *Journal of Wind Energy* **2013** 903057 DOI: 10.1155/2013/903057
[3] Lamedica R *et al* 1997 *IEEE Transactions on Power Delivery* **12(1)** 418–425
[4] Antonio M M 2007 *Power Quality: Mitigation Technologies in a Distributed Environment* (London: Springer-Verlag)
[5] Santoso S *et al* 2012 *Electric Power Systems Quality* (New York: McGraw Hill)
[6] Fuchs E *and* Masoum M 2008 *Power Quality in Power Systems and Electrical Machines*. Edition (London: Elsevier Academic Press)
[7] IEC 61000-3-2 (International Standard) 1995 *Electromagnetic compatibility of technical equipment. Harmonic current emissions (equipment input current ≤ 16 A per phase). Limits and test methods* Retrieved from https://webstore.iec.ch/publication/4149
[8] Desmet J *et al* 2008 *Proc. of the Int. Conf. PESC'08* 785–790 DOI: 10.1109/PESC.2008.4592025
[9] Zhezhelenko I *and* Saenko Y 2007 *Selected problems of non-sinusoidal regimes in electrical networks of enterprises* (Moscow: Energoatomizdat) (in Russian)
[10] *About Software Elcut* 2014 Retrieved from http://www. quickfield.com
[11] *Cables and wires. Power cables* 2014 Retrieved from http://www.sevcable.ru/catalog/mark?id=1119&categoryId=294&industryId=327 (in Russian)
[12] IEC 60287-2-1 (International Standard) 1994 *Electric cables - Calculation of the current rating - Part 2-1: Thermal resistance - Calculation of thermal resistance* Retrieved from https://webstore.iec.ch/publication/1272

IOP Conf. Series: Materials Science and Engineering **189** (2017) 012029 doi:10.1088/1757-899X/189/1/012029

Risk Assessment for Stonecutting Enterprises

A J Aleksandrova[1], and S S Timofeeva[2]
[1]Master, National Research Irkutsk Technical University, Irkutsk, Russia.
[2]Professor, National Research Irkutsk Technical University, Irkutsk, Russia.

E-mail: aleksandrova.angelina1993@yandex.ru

Abstract. Working conditions at enterprises and artisanal workshops for the processing of jewelry and ornamental stones were considered. The main stages of the technological process for processing of stone raw materials were shown; dangerous processes in the extraction of stone and its processing were identified. The characteristic of harmful and dangerous production factors affecting stonecutters is given. It was revealed that the most dangerous are the increased level of noise and vibration, as well as chemical reagents. The results of a special assessment of the working conditions of stone-cutting plant workers are studied. Professions with high professional risk were identified; an analysis of occupational risks and occupational injuries was carried out. Risk assessment was produced by several methods; professions with high and medium risk indicators were identified by results of the evaluation. The application of risk assessment methods was given the possibility to justify rational measures reducing risks to the lowest possible level. The received quantitative indicators of risk of workers of the stone-cutting enterprises are the result of this work.

1. Introduction

Our humanity has been aware of natural stones for a long time. They have been connected with everyday life of human beings since the very first stages of human civilization development. There are good grounds for a prolonged and ancient stage of the humanity history to have been called the Stone Age. At those times people appreciated the role of stones and started using them to make tools, primitive household articles and arms. In the process of mineral properties perception, selection and dressing the humanity accumulated information about stones. Stones were specific in their hardness, colour specifications and the form of appearance. Natural stone had been subjected to a long period of development before it was discovered in carving forms which they were supplied with foremen lapidary. At present, the development of carving art has been extended to different corners of the world, the traditions of stone dressing keeping on their development and opening new orientations.

The art of stone-carving is one of the most ancient expressions of art. Stone products made of ornamental stones are products intended for dressing the interior of living rooms, executive function offices, institutions, etc. The ornaments of stone can beautify a human being as semi-precious stone products also belong to jewellery [1]. For clarity, stones can be divided into practical – decorative ones covering the objects of everyday life, writing sets, pen stands, ashtrays, jewel-boxes, desk-lamps and jewellery and purely decorative ones covering vases, pictures of stone, mosaics, indoor sculpture, etc. [2]. In the interior, the products of stone are most often exhibited on special mounts, display facilities and supports. They can also be displayed on book-shelves, writing tables and as decorations and possess a certain decorative function.

IOP Conf. Series: Materials Science and Engineering **189** (2017) 012029 doi:10.1088/1757-899X/189/1/012029

At present there is a great number of enterprises and craft shops engaged into gem dressing on the territory of the Russian Federation. The enterprises famous for being large in the European part of Russia are the open joint-stock companies of "Moscow Stone Dressing Industrial Complex", "Uralkamen" in Chelyabinsk region in the Urals, "Kolyvan Stone Carving Works" in the region of Altai and the open joint stock company of "Baikalkvarts'samotsvety" in Irkutsk region. Each of these plants started its activities as a small shop where each of the products was carved in a manual way. The process of workpieces production from natural stones is quite labour intensive and complicated. All the technological processes in the above field are inherent in great labour inputs.

2. The aspect of technologies

To find out the way that a stone turns into a subject of admiration one should give consideration to the technological aspect of items production from the stone and understand the risks and problems occurring in the process.

As a matter of fact, searching and prospecting of a stone deposit is the starting stage of exploration. The extraction of gem stones calls for substantially different technological processes and equipment as compared with the extraction of other mineral resources and needs an absolutely different execution order and sequence of mining operations within an open-pit field [3]. Some special variations of stripping and development systems are determined by the above difference. Extraction of the stone is carried out mainly as a result of open-pit operations. Drilling and blasting operations are used to develop stripping [4]. The process of stone raw material extraction substantially affects the enterprise engaged people and is negative for their health.

The open-cut mining of mineral resources is inherent in the following dangers:

- explosive deflagration resulting in no explosion;
- unauthorized explosions during blasting operations en masse;
- water breakthrough or mined rock watering of surface opencast workings, overflowing the area of operations and equipment location;
- collapse of mined rock and steps;
- workers being traumatized and bruised along with getting other somatic injuries when they happen to be trapped in a zone of devices in motion, and some other dangers.

After the extraction, stone blocks are delivered to a stone cut production. This is followed by four production stages: saw cutting, generation of geometry, debugging and polishing.

The technological process of gem items production at a stone dressing enterprise consists of the following operations:

1) Saw cutting is a process of cutting great stone boulders into smaller pieces. Boulders being delivered to the works from a stone deposit cannot be used for direct processing if they are not cut into rather small pieces. To prepare the raw material for dressing stonecutters use different sorts of diamond impregnated saws. They fix them on a corresponding machine-tool with the construction inherent in the sizes of the stones being cut and the specific purpose of cutting. The greatest boulders are cut into thin plates or blocks which are followed by the blocks being cut into smaller work pieces and undercuts (Figure 1a). Precious material is cut with small saws to minimize the wastes [5].

2) Generation of geometry. The process of geometry generation results in the work pieces acquiring their final forms following the surface dressing. Different methods can be used to provide the work pieces with final forms depending on the kind of raw material subjected to dressing and what kind of end product is to be resulting. Each stone is peculiar in its own way, so a special approach is to be used with regard to each one in the process of geometry generation (Figure 1b). At the given stage the product acquires its primary outlook of a designed item with the assistance of such instruments as drilling tools, abrasive tools, miniature cutting saws, manual abrasives and other auxiliary devices. On imparting a sketchy outlook to a product it is directed to be debugged.

3) Debugging. The operation being considered is an intermediate one between the generation of geometry and polishing. It is the last stage of stone dressing in the process of geometry generating prior polishing. The debugging process is not complicated: all you need is to rub the stone against a

flat plate having some abrasive material put on it. This mode is always applied for when an even surface is to be obtained on a flat cut or on thin parts of a product (Figure 2). Debugging usually starts from using some coarse grain abrasive and is finished by using the finest of it [6].

4) Polishing. The final dressing of a product covers making a stone surface fully brilliant and smooth [7]. The quality of this kind stone dressing depends on the previous stages of geometry generation and debugging. Dazzling and smooth surfaces that are usual for polished stones result from the surface contact with a rotating soft material such as felt, leather, fabric or wood (Figure 2b). The materials are loaded with polishing pastes and contribute to stone polishing. Sticks made of hard wood can be used to dress hard-to-reach spots. They are fixed in a collet holder of a drilling tool and overdone with the paste enriched in a micropowder.

a) b)

Figure 1. The operations of stone-cutting production, where a is the process of stone cutting, b is the process of formation.

3. Risk analysis

Professional risks occur at each stage of stone material dressing. They are a number of physical, chemical, psychophysical factors that are dangerous and impregnated with a harmful effect for human health. The workers are exposed to such dangers as those of moving products and semiproducts of stones, dust in the process of cutting and polishing the stone, the temperature of the stone surfaces being dressed, moving devices and mechanisms, cutting waste from metallic tools cutting the stone, etc. Modern output of stonecutting products is characteristic for the use of equipment with dangerous technological parameters in the processing [8] (noise, vibration, air dustiness, moving tools) and a sufficient level of different chemicals.

A number of professions characterized by high professional risks were revealed on investigating and analyzing the resulting data of special estimating the working conditions of workers for stone-cutting productions and a shop engaged in raw gemstones dressing which is a production shop of jewellery. The professions of this kind are as follows: a grinder, a polisher of stone products, a stone cutter, a jeweller-assembler, a jeweler – integrator, a carver of stones [9].

Several methods were used to assess professional risks for most dangerous professions: predictive assessment of professional risks, determination of an individual professional risk for a worker, evaluation of professional risks on the official place of work resulting from questionnaire. The methods of predictive assessment and determination of an individual professional risk are objective ones as they are based on the results of official place of work assessment performed at an enterprise, whereas professional risk assessment on the official place of work based on questionnaire is a subjective one as it results from workers interrogation. The results of professional risk assessment based on these methods revealed that every worker of a stone cutting production has to be employed in some operation environment conditions which make a negative effect on their health.

a

b

Figure 2. The operations of stone-cutting production, where a is the process of polishing stone; b is stone polishing process.

With regard to safety, stone cutting industry is one of the dangerous productions [10] and leads to feedback effects influencing the workers' health in the course of performing technological operations. It means that before we admire the art of stone carving a great number of complicated stages and operations are to be completed to extract and dress stone material.

The present paper deals with the assessment of professional risks resulting from stone dressing and jewelry making having taken the open joint-stock company of "Baikalkvarts'samotsvety" located in Irkutsk region of the Russian Federation as an example.

Unsalutary production factors were revealed as a result of instrumental measurements. The used measuring instruments were the ones that had been exposed to a state control and their metrological data were certified in the prescribed procedure.

The presence of steady sources of an intense noise was found to be a specific peculiarity of the stone-dressing industry, manufacture of carved stone products including: production machinery, stone carving tools (the sound of hydro-abrasive machinery and ultrasound found in the operation environment, vibration perceived at every stage of production), fans and exhaust plants.

Noise measurements were performed in accordance with the authorized test procedure [11]. The results obtained are presented in Table 1.

Table 1. The results of noise level evaluation on the official work places.

The indication of a work place	MAL, [dBA]	Real level of work environment factor, [dBA]	Working conditions class
Grinder-polisher	70	74.4	3.1
Stone sawer	70	74.4	3.1
Jeweller-assembler	70	74.4	3.1
Jeweller-integrator	70	74.4	3.1
Carver of stone	70	74.4	3.1

As follows from the data represented in Table 1, the indices of noise level on all the official work places exceed authorized level limits. This is indicative of the fact that the labor condition class for noise level at these places is harmful (Class 3.1). The excess of noise level at the official work places is conditioned by the workers being in a close proximity to the production machinery which is a source of noise affecting the functional state of a worker's organism.

Chemicals in the production of stone-cut workpieces are represented as harmful substances and belong to 1-4 Classes of danger. They are characterized by having different effects on the organism, some of them possessing a delayed and slow aftereffect. To calculate the concentration the methodology of photometric makeup estimation was used for each of the substances available in the work zone air [12].

The production workers of such professions as jeweler-assembler, stuff-setter, polisher are commonly exposed to chemicals effect.

Table 2 shows the readings of chemicals in the work zone air of the professions being investigated.

Table 2. Air pollution of the work environment with chemical compounds in the stone cutting production.

Indication of chemicals (in the work zone)	Real value	Standard value	Danger class	Working conditions class	Exposure time
Phthalic anhydride, mg/m^3	0.29	1	2	2	70
Acetone, mg/m^3	15	200	4	2	60
Formaldehyde, mg/m^3	0.012	0.5	2	2	50
Hexamethylendiamine,mg/m^3	0.71	1	2	2	85
Time weighted averages of concentrations (TWA)					**100**
Phthalic anhydride, mg/m^3	0.07	1		2	
Acetone, mg/m^3	13	200		2	
Formaldehyde, mg/m^3	0.007	0.5		2	
Hexamethylendiamine,mg/m^3	0.64	1		2	

Basing on the results of experimental measurements a calculation of professional risks was performed. It was found that the risk tended to be the medium one. Besides, special attention should be paid to psychophysiological factors of production, such as hearing and eyesight tension, considerable physical loading, limited, single-type movements of hands and special sets of fingers, prolonged, tense, inconvenient working posture, steady loads.

4. Conclusion

A stone-cutting industry is a potentially dangerous field for the workers engaged. A group of unfavourable factors characteristic for the operational environment (vibration, noise, a chemical factor) is formed by peculiarities of the technological process, the equipment used in it, mechanization and automatization levels of stone-cutting workpieces production.

In spite of the fact of a negative effect in the working conditions of stone-cutting production being proved the number of its workers does not become less. From the data obtained in literature it is clear that the production under consideration belongs to the category of dangerous production objects resulting in the number of people exposed to professional, accidental and ecological risk growing from year to year.

That was the reason of assessing the risks of disability for workers engaged in the stone-cutting production, the state of working conditions being paid special attention to as they exert influence on formatting professional risks, criteria of industrial security and the development of possible events oriented at risk minimization. As a result of the assessment, noise and chemicals were revealed to be the most unfavourable factors of risk.

The probability of a professional risk realization can be found at any industrial object. Therefore, the application of the methodology assessing the risk of dangers allows to reason some rational measures that allow to make the risks as low as possible. The methodology of predictive risk assessment can be objective and subjective. The paper under consideration deals with the risk assessment in a stone-cutting production which is the identification of dangers typical for the given field, determination of danger dimensions in regard to a worker's health and their possible after-effects.

Thus, the investigations performed allow to assess the risk for the health of production workers.

References

[1] Alexandrova A Yu *and* Timofeeva S S 2016 *Proc. of "Technosphere Safety in XXI Century"* 24–32 (in Russian)
[2] Jerrard R *et al* 2007 *Design management: Exploring fieldwork and applications* **5** 102–121 (London: Routledge)
[3] Bordunov S V *and* Galtseva O V 2016 *J. Phys.: Conf. Series* **671(1)** 012009 DOI: 10.1088/1742-6596/671/1/012009
[4] *Mining encyclopedia* Retrieved from http://www.mining-enc.ru (in Russian)
[5] Robleda A S *et al* 2010 *DYNE (Colombia)* **77(161)** 77–87 ISSN: 00127353
[6] Konstanty J 2002 *J. Mater. Process. Technol.* **123(1)** 146–154 DOI: 10.1016/S0924-0136(02)00071-7
[7] Sinkankas J 1989 *Guidelines for the Processing of Precious and Semiprecious Stones* (Moscow: Mir) (in Russian)
[8] Dorofee A J *et al* 1996 *Continuous Risk Management Guidebook* (Pittsburgh, PA: Software Engineering Institute, Carnegie Mellon University)
[9] Alexandrova A Yu 2016 *Proc. of "Security Problems of the modern World"* 7–8 (in Russian)
[10] *Encyclopedia of Safety* Retrieved from http://survincity.com/2012/04/stone-cutting-art-cultural-richness-of-the-country/
[11] GOST (State Standard) R ISO 9612-2013 *Acoustics. Noise measurements for the evaluation of its effects on humans. Measuring method workplace* (Moscow: Standartinform) (in Russian)
[12] *Chemical Encyclopedia* Retrieved from http://www.xumuk.ru/encyklopedia/ (in Russian)

IOP Conf. Series: Materials Science and Engineering **189** (2017) 012030 doi:10.1088/1757-899X/189/1/012030

Soil ph Control in the Mobile Corrosion Monitoring

T A Radenkov[1], S V Romanenko[2], V A Kolpakov[3], O E Merzlyakov[4], V A Kabanov[5], and A G Kagirov[6]

[1]Assistant, Tomsk Polytechnic University, Tomsk, Russia
[2]Professor, Tomsk Polytechnic University, Tomsk, Russia
[3]Engineer-Chemist, Tomsk Polytechnic University, Tomsk, Russia
[4]Associate Professor, Tomsk State University, Tomsk, Russia
[5]Student, Tomsk Polytechnic University, Tomsk, Russia
[6]Lecturer, Tomsk Polytechnic University, Tomsk, Russia

E-mail: drovosek@tpu.ru

Abstract. Corrosion monitoring methods and pH measuring devices are studied. Considering the heterogeneity of the soil pH, depending on depth and sampling points, it is advisable to use solid-state potentiometric sensors that are capable of pH measuring with an error not exceeding 0.2 pH units. In the process of the pipeline installation necessary to control the soil pH value. While it is impossible to predict what kind of horizon will be in contact with the pipe material at different points. This problem requires future development of technique for the correct pH measuring for the purpose of mobile pipeline corrosion monitoring. When using cathodic protection has been required find the balance between the parameters which prevent the corrosion reactions and also take into account the hydrogen absorption on surface of the pipes material.

1. Introduction

Currently, pipeline transport one of most used method of energy transportation. Important effect accompanying the pipelines operation is the pipe material corrosion. Most emergencies occurring with pipelines exploitation are associated with corrosion, which is one of the main causes of depressurization.

Nowadays, corrosion monitoring is an effective way to estimate the equipment condition (in the stages of its design, operation, renovation). Monitoring of the underground steel pipelines corrosion state includes a system of observations, diagnosis and corrosion condition prediction for timely defect detection, estimating their extent and prevent of the effects of corrosion.

In corrosion monitoring at the stage of equipment operation using the following methods condition monitoring: visual inspection, determination of the medium redox potential, examination of inaccessible areas of equipment with telemetry systems, monitoring changes in the concentration of corrosion-active agents, etc., degradation products of the metal structure elements, determining material construction potential, determination of corrosion specimens kinetics, determination of specimens electrical resistance, ultrasonic, acoustic and magnetometric defectoscopy [1].

In mobile corrosion monitoring there is a problem consisting in sampling and subsequent analysis in the laboratory. This leads to high costs of time and human resources.

2. Research Questions

There are passive and active methods of protection material of pipelines from corrosion. Because the passive method doesn't provide complete corrosion protection of pipelines, use an active protection based on the electrochemical processes occurring at the border of the soil electrolyte and pipe metal. The most common active method of corrosion protection is the cathodic polarization method based on the shift of pipe metal potential in the negative region relative to the equilibrium potential therefore the rate of metal dissolution is decreased [2].

The intensity of the pipeline material destruction considerably depends on the soil corrosiveness in contact with the pipe. Soil corrosivity is determined by the following parameters: moisture, permeability of soil, porosity, gas-phase composition of its pore space, organic compounds concentration in soil, acid and sulfate-reducing microorganisms, as well as pH, salinity and mineral salts composition of the soil electrolyte [3]. Selection of pipeline protection method directly based on the parameters of soils corrosion activity.

In determining the soil corrosiveness there is an effective tool, displays the thermodynamically stable form of elements existence (atomic crystals, metals, molecules and ions) in the soil solutions at various pH and the redox potential E. Proposed by Marcel Pourbaix [4], the chart is plotted in coordinates of E (ordinate) — pH (abscissa)(Fig. 1). It reflects the thermodynamically stable Fe form at a corresponding pH value and redox potential. For each chemical element, it is possible to design a Pourbaix diagram.

The Pourbaix diagram is a useful tool in determining the possible corrosion reactions on the pipeline, and also in the potential of cathodic protection determining.

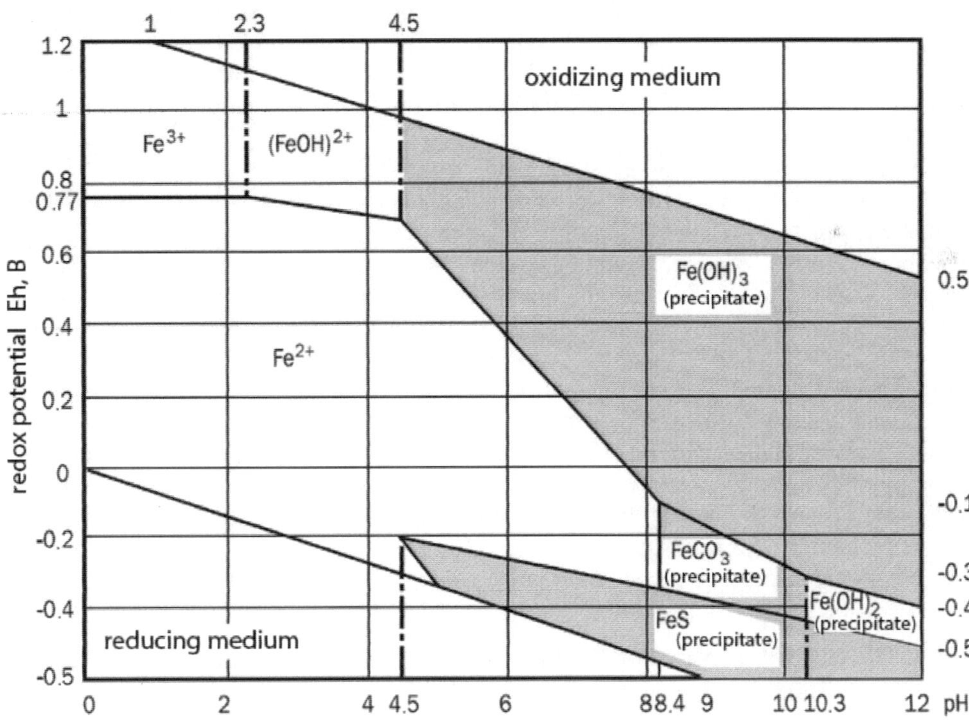

Figure 1. The possible reactions for the Fe-H_2O system.

Determination of soils pH is one of the important elements of soils corrosion activity evaluation. pH quantitatively expresses a solution acidity. It is a negative decimal logarithm of hydrogen ion concentration (exactly – activity). The soil is a multifunctional heterogeneous open disperse system formed by the weathering of rocks and living organisms. The soil profile is a set of horizons, which differ in chemical, mineralogical, granulometric composition, physical and biological properties. Soils are divided into types that differ from each other by a set of properties caused by the processes of soil formation and the major horizons system [5–7].

3. Research Methods and Problem Statement

Thus, it is necessary to develop equipment for measuring soil pH is suitable for use in mobile corrosion monitoring. Among the existing pH measuring methods, the most common method is the potentiometry. This method has sufficient accuracy for the pH determination (to 0.01 pH unit), small measurement time (about three minutes) and sufficient results reproducibility. In practice, the measuring system consists of two electrodes, one of which is sensitive to the hydrogen ions concentration, and the other one has a stable potential (the reference electrode). As a pH sensor usually used national standard recommended [8] glass electrode, and silver-chloride electrode as reference one.

Despite the high accuracy of the glass electrode [9] has a number of disadvantages:

- fragility;
- long-establish the equilibrium potential value;
- high electrode resistance;
- special storage conditions and preparation to work.

In addition to the glass electrode for soil pH measuring used antimony electrode [7]. It has a lower accuracy in comparison with a glass electrode, but is more suitable for use at-site.

In the literature [10, 11] it is known to use solid-state sensors based on quinhydrone to measure the pH of the environmental objects. Such solid-state electrodes have the following advantages:

- strength, adequate for use in field conditions;
- potential stabilization within minute;
- storage , transportation and preparation for work without specific conditions;
- low electrode resistance;
- wide sensor shape and size variation;

There are various methods for soil pH determininating characterized by different measurement errors. It is known from the literature [12] that the range of pH variation within a single soil type varies within 0.5–1 pH. Also, depending on the measurement depth, pH changes in each soil horizon from top to bottom from acidic to alkaline (from 4.0 in swampy soil types to 8.5 in pedogenic horizons).

4. Results and Findings

As a result of the experiment was determined variation of soil pH values sampled in the Tomsk and Kemerovo regions, depending on soil type, depth of sampling point in accordance with the methodology of GOST 26423-85.

Soil profile — a set of genetically mated soil horizons, which divided the soil in the process of soil formation [13]. Table 1, 2 illustrate the pH variation in different soil types and at different horizons. The following soil types: podzolic, light gray forest, alluvial meadow acidic, turfy, light brown, ordinary black earth forest-steppe, black humus, sod.

Table 1. The range of pH variation for different soil types (degree of freedom 3, significance level 0.05).

	Black humus	Podzolic	Turfy	Black earth	black earth forest-steppe
pH	6.05 ± 0.16	5.73 ± 0.19	6.33 ± 0.15	7.52 ± 0.3	7.29 ± 0.19
Confidence interval	0.32	0.38	0.3	0.6	0.38

These tables show that relation on soil type pH values spreading up to 0.6 pH units.

Table 2 illustrates pH variation in a soil horizons in different types of soils.

The differences in pH values between the various horizons are statistically significant and reaches 1 pH. Soil horizons are mixed when a pipeline is laying. Thus, the pipe contact with soil with different pH value.

Table 2. The range of soil pH variation on different horizons (degree of freedom 3, significance level 0.05).

	Podzolic	Dark humus	Black Earth	Turfy
pH	A: 5.51 ± 0.28	A: 5.74 ± 0.31	A: 7.19 ± 0.32	A: 5.83 ± 0.22
	A2B: 5.42 ± 0.24	A1A2: 5.98 ± 0.30	AB: 7.32 ± 0.24	A1A2: 6.55 ± 0.27
	B1: 6.4 ± 0.22	B1: 6.52 ± 0.27	B1: 7.88 ± 17	B1: 6.7 ± 0.24
pH range	5.51 ÷ 6.4	5.74 ÷ 6.52	7.19 ÷ 7.88	5.83 ÷ 6.7

To determine the pH sampling point influence was selected 3 areas of soil in Mikhaylovskaya grove (location - Tomsk city, soil type – light-gray forest), distance between sampling points approximately 5 meters.

Table 3. Variability of pH values in different sampling points within the same soil type (degree of freedom 3, significance level 0.05).

	1	2	3
pH	7.3±0.2	7.8±0.3	8.2±0.3
Confidence interval	0.4	0.6	0.6

From Table 3, we can conclude that depending on the selection point within the same soil type, the pH value can vary almost 1 pH unit.

5. Conclusions

Considering the heterogeneity of the soil pH, depending on depth and sampling points, it is advisable to use solid-state potentiometric sensors that are capable of pH measuring with an error not exceeding 0.2 pH units.

In the process of the pipeline installation necessary to control the soil pH value. While it is impossible to predict what kind of horizon will be in contact with the pipe material at different points. This problem requires future development of technique for the correct pH measuring for the purpose of mobile pipeline corrosion monitoring. When using cathodic protection has been required find the balance between the parameters which prevent the corrosion reactions and also take into account the hydrogen absorption on surface of the pipes material.

Solid-state sensors for the measurement of pH for corrosion monitoring are effective and have the following properties:

- measurement error less than 0.2 pH units;
- the range of pH 4 to 8;
- easement of using at site;
- mechanical durability.

References

[1] Saji V S *and* Cook R 2012 *Corrosion Protection and Control Using Nanomaterials* (UK: Woodhead Publishing Ltd.)
[2] Groysman A 2010 *Corrosion for Everybody* (Netherlands: Springer)
[3] Pesterfield L L *et al* 2012 *Journal of Chemical Education* **89(7)** 891–899 doi: 10.1021/ed200423n
[4] Sanders M H *and* Phillip D C 2013 *Geomorphology: Processes, Taxonomy and Applications* (US: Nova Science Publishers)
[5] Orazem M E 2014 *Underground Pipeline Corrosion: Detection, Analysis and Prevention*, (UK: Woodhead Publishing Ltd.)
[6] Yurchenko A V *et al* 2016 *IOP Conf. Ser.: Mater. Sci. Eng.* **671** 012001 DOI: 10.1088/1742-6596/671/1/012001
[7] Yakunin M A *and* Yurchenko A V 2015 *Tech. Phys.* 60(1) 141–144 DOI: 10.1134/S1063784215010272

[8] GOST (State Standard) 26423-85 *Soils. Methods for determination of conductivity, pH, and the solid residue of the aqueous extract* (Moscow: Standartinform) (in Russian)
[9] Bates G R 1964 *Determination of pH. Theory and practice* (UK: John Wiley & Sons)
[10] Radenkov T *et al* 2011 *"Testing. Diagnostics"* **146** 146–148 (in Russian)
[11] Scholz F *et al* 1993 *J. Anal. Chem.* **347(10)** 458–459 doi:10.1007/BF00635475
[12] Pansu M *and* Gautheyrou J 2006 *Handbook of soil analysis: Mineralogical, organic and inorganic methods* (Berlin: Springer Berlin Heidelberg)
[13] GOST (State Standard) 27593-88 *Soil. Terms and Definitions* (Moscow: Standartinform) (in Russian)

IOP Conf. Series: Materials Science and Engineering **189** (2017) 012031 doi:10.1088/1757-899X/189/1/012031

Structural and Phase Changes in the System Al-Si-Ti-B, Synthesized Using the Electron-Ion-Plasma Treatment Method

A A Klopotov[1, 2], E A Petrikova[3], Yu F Ivanov[1, 3], A D Teresov[3], N N Cherenda[4], V V Uglov[4], and N A Tsvetkov[2]

[1]Professor, National Research Tomsk State University, Tomsk, Russia
[2]Professor, Tomsk State University of Architecture and Building, Tomsk, Russia
[3]Junior Researcher, Institute of High-Current Electronics, Siberian Branch of Russian Academy of Sciences, Tomsk, Russia
[4]Professor, Belarusian State University, Minsk, Republic of Belarus

E-mail: klopotovaa@tsuab.ru

Abstract. The results of surface modification of silumin samples of the eutectic composition using the combined method, which includes the irradiation with plasma formed during an electric explosion of the conductive material and the subsequent processing with a high-intensity pulsed electron beam, are presented. Formation of a multilayer multiphase submicron and nanoscale structure with high mechanical and tribological properties is established. The results of the structural phase analysis of the material are discussed from the standpoint of thermodynamics.

1. Introduction

Pulsed melting with the simultaneous saturation of surface layers of the material with doping elements and the subsequent crystallization and formation of strengthening phases, carried out by plasma formed during an electric explosion of the conductive material (electroexplosive doping), is one of the promising methods in the modification of the structure and properties of metals and alloys [1]. The formed layers have a metallurgical bond with the substrate having the adhesion at the level of cohesion and slightly change dimensions of the item. An additional increase in properties of the surface layer of the material and the item is generally possible during the subsequent heat treatment [2]. An effective method of heat treatment is high-intensity pulsed electron beams [3]. As compared to the widespread laser technology, the electron-beam technology has great potential in monitoring and control of the amount of the input energy, differs in the locality of energy distribution in the near-surface layer of the treated material and a high efficiency coefficient. Ultra-high rates ($10^6 \div 10^8$ K/s) of heating up to melting temperatures and the subsequent cooling of a thin near-surface layer ($10^{-7} \div 10^{-6}$ m) of the material, a very small time of exposure to high temperatures ($10^{-6} \div 10^{-3}$ s), formation of limiting temperature gradients (up to 10^8 K/m) that provide cooling of the near-surface layer due to heat transfer into the integrally cold volume of the material at a rate of up to 10^7 K/s, create conditions for formation of an amorphous and nanocrystalline structure in the near-surface layer [4].

The aim of the present work is to reveal evolution patterns of the structure and the phase composition of the alloy aluminum-silicon (silumin) subjected to electroexplosive doping and the subsequent irradiation with a high-intensity pulsed electron beam of a submillisecond exposure time.

2. Material and methods of research

The alloy Al–12.5% Si (eutectic silumin) was selected as a research material [5]. Electroexplosive doping of the silumin surface was carried out on the installation EVU 60/10 (Siberian State Industrial University, Novokuznetsk). The following mode of doping with plasma formed during an electric explosion of titanium foil with a sample of boron powder was used: the exposure time of plasma is ~ 100 μs, the absorbed power density on the axis of the jet is ~ 8.2 GW/m^2, the pressure in the shock-compressed layer near the surface is ~ 18.8 MPa; the thickness and the weight of titanium foil is about ~ 20 microns, ~ 90 mg, the weight of boron powder is ~ 90 mg. The subsequent melting of the modified surface layer of silumin was carried out with a high-intensity pulsed electron beam on the installation "SOLO" (IHCE SB RAS) [6] under the mode: the electron energy is 18 keV, the energy density of the electron beam is (15÷30) J/cm^2, the pulse duration is 200 μs, the number of pulses is 5, the pulse repetition rate is 0.3 s^{-1}; the irradiation was performed in argon at a residual pressure of 0.02 Pa. The following took place at the given parameters of the electron beam: heating of the near-surface layer of silumin with a thickness of ~ 30 microns up to melting temperature, crystallization and quenching from the melt at rates up to 10^6 K/s. The study of the structure of the modified layer of silumin was carried out using methods of metallography, scanning and transmission diffraction electron microscopy, X-ray diffraction analysis.

3. Results and discussion

The combined treatment of silumin surface, which includes the irradiation with plasma formed during an electrical explosion of the conductive material and the subsequent irradiation with a high-intensity pulsed electron beam, is accompanied by a multiple increase in the strength and tribological properties of the near-surface layer of the material. Analyzing the results obtained during tests it is possible to note the following. First, the combined treatment, which includes doping with plasma of an electric explosion of titanium foil with a sample of boron powder and the subsequent irradiation with an electron beam, is accompanied by hardening of the surface layer of silumin up to the value 7.5 times greater than the microhardness value of the sample core. Secondly, the combined treatment results in formation of extended hardened layers, the thickness of which reaches 200 μs. Thirdly, the microhardness and the thickness of the hardened layer of silumin depend on the mode of the subsequent irradiation with an electron beam. Namely, an increase in the energy density of the electron beam (at constant values of other parameters of the beam) leads, as a rule, to an increase in the microhardness and the thickness of the hardened layer.

High level of hardness of the surface layer of silumin formed during the combined treatment, which includes electroexplosive doping and the subsequent irradiation with an electron beam, leads to a significant decrease in the friction coefficient of the material. During an electroexplosive doping of silumin with plasma formed suring an electric explosion of titanium foil with a sample of boron powder and the subsequent irradiation with an electron beam with parameters of 25 J/cm^2, 200 ms, 5 pulses, 0.3 s^{-1} a decrease in the friction coefficient in 5-6 times is observed (counterbody – solid alloy VK8, the load on the indenter is 20 g). It shall be noted that at the given treatment mode the sample of silumin was characterized by an extended modified layer with a high, as compared to the initial material, microhardness level.

Thus, the carried out studies show that the combined treatment, which includes doping of the surface layer of silumin with plasma formed during an electric explosion of the conductive material and the subsequent irradiation with a high-intensity electron beam, allows to modify silumin in the volume with a thickness of up to 200 microns; surface layers are characterized by high level of microhardness and low values of the friction coefficient. It shall be expected that hardening mechanisms of the surface layer of silumin, implemented using the given treatment method, are: solid solution (formation of aluminum-based solid solutions), dispersion (formation of nanoscale particles of second phases), grain boundary (refinement of the grain structure), and deformation (formation of the defect substructure).

It is obvious that the main of the above mentioned hardening mechanisms in the investigated alloy is dispersion, caused by formation of nanoscale particles of the second phase. Fig. 1 presents isothermal sections of ternary systems Al–Si–B [8], Al–Ti–Si [9], Al–Ti–B [10], Ti–Si–B [11] demonstrating many phases with an ability to form in the investigated alloy under equilibrium conditions. An important point for the considered alloy is the fact that, on the one hand, borides and silicides of transition metals, in regards to a range of physical-and-mechanical properties, are attributed to metal compounds with high electrical conductivity having temperature dependences of the thermal electromotive force in pairs with metals, similar to metal pairs [12, 13]. On the other hand, silicides and borides of titanium cannot be attributed to compounds of the implementation phase type since they do not satisfy the Hagg's condition and considerably exceed the Hagg's critical ratio of 0.59. This, for example, leads to the fact that in binary systems Ti–Si the substitution with silicon atoms is accompanied by formation of metal structures at a ratio (R_{Si}/R_{Me}) >0.9. The content of silicon in silicides can reach values of about 50 at%. At higher contents of silicon, the substitution of metal atoms with silicon leads to formation of complex crystalline structures and to an increase in the role of the covalent bond among silicon atoms [14] (Figure 1).

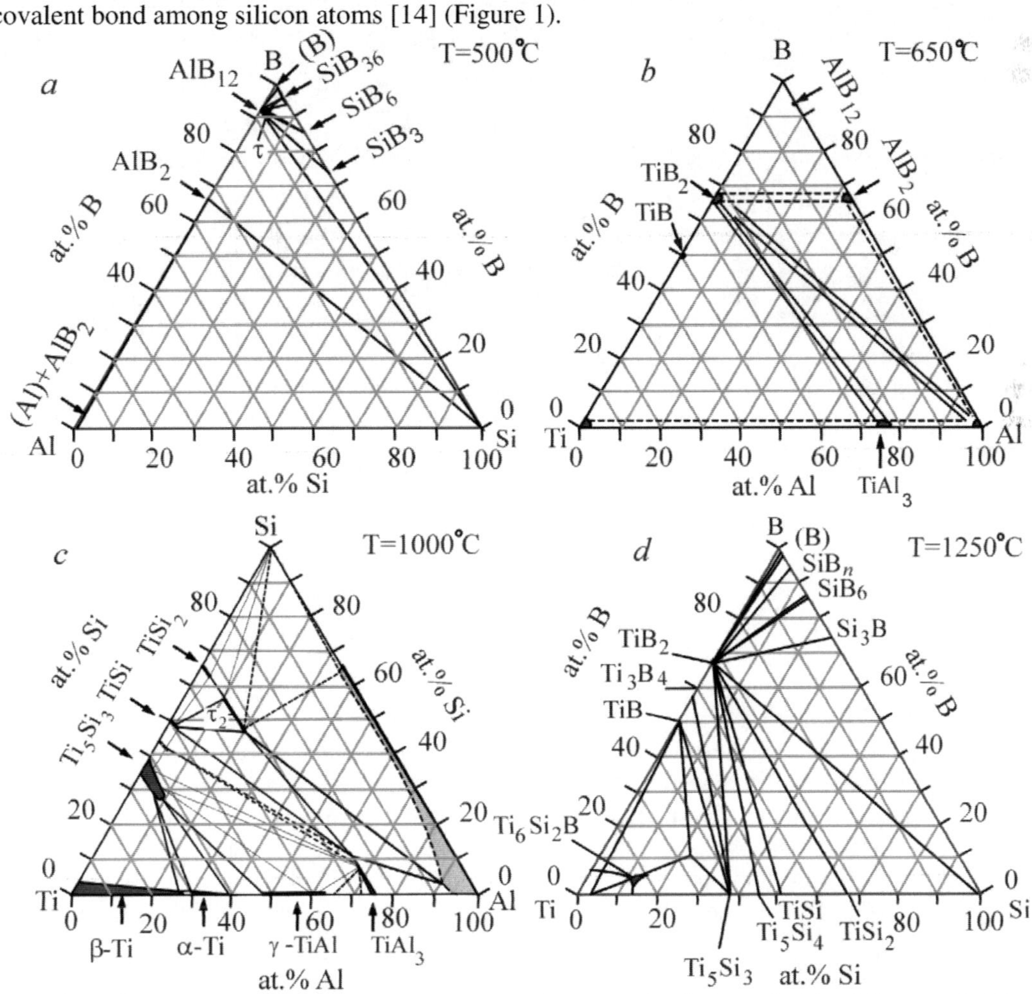

Figure 1. Isothermal sections of ternary systems, where *a* is Al–Si-B at 500 °C [8]; *b* is Al–Ti–Si at 1000 °C [9]; *c* is Al–Ti–B at 650 °C [10]; *d* is Ti–Si–B at 1250 °C [11].

IOP Conf. Series: Materials Science and Engineering **189** (2017) 012031 doi:10.1088/1757-899X/189/1/012031

Treatment of silumin surface with plasma formed during an electric explosion of a conductive material leads to formation of the doped layer with a thickness of tens of micrometers and a high level of roughness, containing a large amount of microcraters, cracks, particles and/or fragments of the exploded material, characterized by a high level of heterogeneity of the surface layer by alloying elements (Figure 2a).

It is obvious that deficiencies do not allow the direct use of the electroexplosive doping method for treatment of silumin items. In the present work, as mentioned above, the additional treatment of the modified surface of silumin was carried out by an intense pulsed electron beam. The electron-beam treatment of samples subjected to electroexplosive doping leads to a significant change in the structure and the elemental and phase composition of the modified layer. Most notably, this change takes place in the morphology and the roughness level of the alloyed surface (Figure 2b).

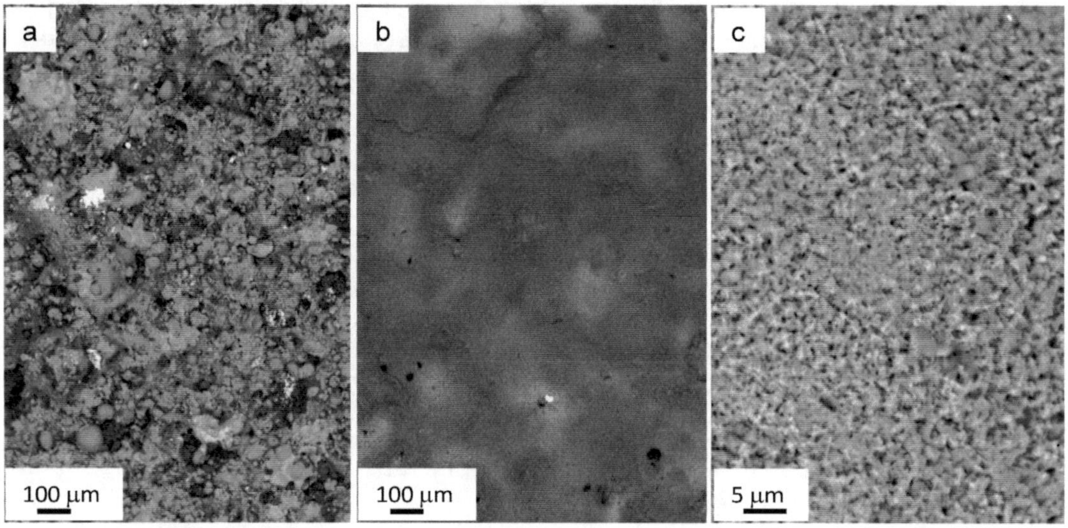

Figure 2. The structure of silumin modification surface, where *a* is with plasma formed during an electric explosion of titanium foil with a sample of boron carbide powder; *b, c* is after the additional irradiation with an intense pulsed electron beam with the parameters of 30 J/cm^2, 200 μs, 5 pulses.

Melting of the surface layer, which occurs when the energy density of the electron beam exceeds 20 J/cm^2, and the subsequent high speed crystallization of the surface layer leads to formation of a structure, the crystallite size of which vary in the range of tens and hundreds of nanometers (Figure 2c).

Treatment of the doped layer of silumin with an electron beam is accompanied by an alignment of its thickness, a decrease in the number of microcracks and micropores (Figures 3a, 3b), a significant grinding of silumin layer adjacent to the modified surface layer (Figures 3c, 3d).

The studies of the phase composition of the surface layer of silumin subjected to a complex treatment, that combines doping with plasma formed during an electric explosion of titanium foil with a sample of boron powder and the subsequent irradiation with an intense electron beam, were carried out using methods of X-ray analysis.

By analyzing the X-rays shown in Figure 4 it can be noted that this doping method (electroexplosive doping and the subsequent electron-beam treatment) leads to formation of the following additional phases in silumin layer (along with Al and Si): Ti; Al_2Ti; $AlTi_3$; Al_3Ti; TiB.

The state of the defect substructure of phases of the surface layer of silumin subjected to electroexplosive doping and the subsequent electron-beam treatment has been analyzed using methods of diffraction electron microscopy by studying thin foils. First, the dispersion effect (up to submicron and nanoscale values) of the structure of main phases of silumin (aluminum and silicon), which agrees well with the results obtained using methods of scanning electron microscopy, shown in Figure 2 and Figure 3. The dimensions of phases formed during doping also vary in a wide range: from nano up to

IOP Conf. Series: Materials Science and Engineering **189** (2017) 012031 doi:10.1088/1757-899X/189/1/012031

submicron. In addition to phases detected using methods of X-ray analysis, nanoscale particles of titanium silicide TiSi have been revealed in the volume and at boundaries of aluminum grains, as well as at interphase boundaries aluminum/silicon.

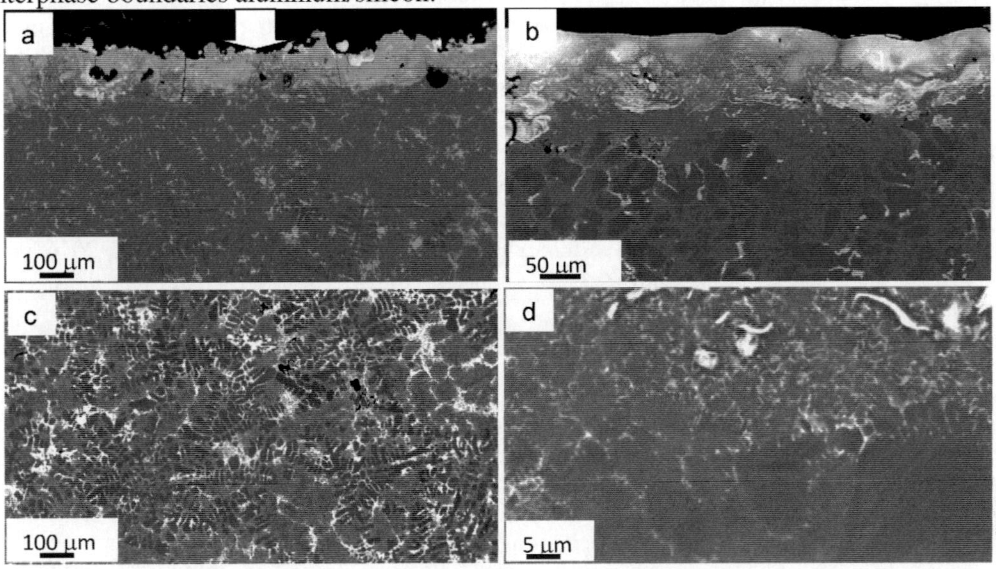

Figure 3. The structure of silumin cross-section is in the initial state (*c*); the structure after doping with plasma formed during an electric explosion of titanium foil with a sample of boron carbide powder (*b*) and after the additional irradiation with an intense pulsed electron beam with the parameters of 30 J/cm^2, 200 µs, 5 pulses. (*a, d*). In (*a*) the arrow indicates the modified surface.

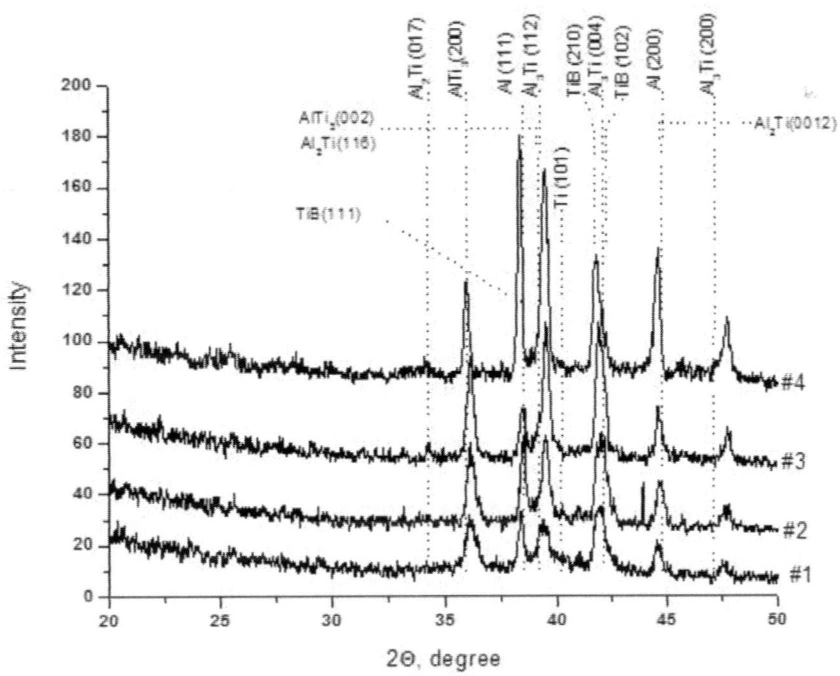

Figure 4. Sections of radiographs of the silumin sample subjected to doping with plasma formed during an electric explosion of titanium foil with a sample of boron powder (1); samples subjected to an additional treatment with an intense pulsed electron beam at various electron beam energy densities (200 µs, 5 pulses. 0.3 c^{-1}): 20 J/cm^2 (2); 25 J/cm^2 (3); 30 J/cm^2 (4).

4. Conclusion

The studies on the combined electron-ion-plasma treatment of silumin surface, which include doping with plasma formed during an electric explosion of titanium foil with a sample of boron powder and the irradiation with a high-intensity pulsed electron beam, have been presented. Formation of a multiphase submicron and a nanoscale structure of the surface layer, characterized by a significant increase in the strength and tribological properties, has been revealed.

Acknowledgement

The research was supported by the Russian Foundation for Basic Research grant (the project No. 16-58-00075 Bel_a) and in the course of the implementation within the framework of the Program "Scientific Foundation named after D.I. Mendeleev of Tomsk State University" in 2015-2016.

References

[1] Gromov V E *et al* 2015 *Fatigue of steels modified by high intensity electron beams* (Cambridge: Cambridge International Science Publishing Ltd)
[2] Karpiy S V *et al* 2010 *The structure, phase composition, and properties of titanium after electroexplosive doping and electron-beam treatment* (Novokuznetsk: NPK) (in Russian)
[3] Devyatkov V N *et al* 2003 *Laser and Particle Beams* **21(2)** 243–248 DOI: 10.1017/S026303460321212X
[4] Rotshtein V *et al* 2006 *Materials surface processing by directed energy techniques* **6** (London: Elsevier) 205–240 ISBN: 9780080444963
[5] Stroganov G B 1977 *Alloys of aluminum with silicon* (Moscow: Metallurgy) (in Russian)
[6] Koval N N *et* 2009 *IEEE Trans. on Plasma Science* **37(10)** 1890–1896 DOI:10.1109/TPS.2009.2023412
[7] Liakishev N P 1966–2000 *State diagrams of binary metallic systems* (Moscow: Mechanical engineering) (in Russian)
[8] Yoshikawa T *and* Morita K 2005 *Mater. Trans* **46(6)** 1335–1339 DOI: (10.2320/matertrans.46.1335)
[9] Brukl C *et al* 1961 *Monatsh. Chem.* **92** 781–788 DOI: 10.1007/BF00918638
[10] Fjellstedt J *and* Jarfors A E W 2001 *Zeitschrift fuer Metallkunde/Materials Research and Advanced Techniques* **92(6)** 563–571 ISSN: 00443093
[11] Candioto K C *et al* 2011 *J. Alloys Compd.* **509(17)** 5263–5268 DOI: 10.1016/j.jallcom.2011.02.042
[12] Samsonov G V 1959 *Silicides and their use in engineering* (Kiev: Ukrainian Academy of Sciences) (in Russian)
[13] Frantsina E V *et al* 2014 *Chemical Engineering Journal* 238 129–139
[14] Serebriakova T I *et al* 1991 *High-temperature borides* (Moscow: Metallurgy) (in Russian)

IOP Conf. Series: Materials Science and Engineering **189** (2017) 012032 doi:10.1088/1757-899X/189/1/012032

Method and Apparatus of Measuring Velocity and Sound Attenuation Coefficient in Bulk Materials Based on the Analysis of the Structure of Sound-Insulation Materials on the Basis of Perlite

B I Kapranov[1], and A P Mashanov[2]

[1]Professor, National Research Tomsk Polytechnic University, Tomsk, Russia
[2]Student, National Research Tomsk Polytechnic University, Tomsk, Russia

E-mail: introbob@mail.ru

Abstract. This paper presents the results of research and describes the apparatus for measuring the acoustic characteristics of bulk materials. Ultrasound, it has passed through a layer of bulk material, is further passes through an air gap. The presence of air gap prevents from measuring tract mechanical contacts, but complicates the measurement technology Studies were conducted on the example of measuring the acoustic characteristics of the widely used perlite-based sound-proofing material.

1. Introduction

Various types of bulk materials widely used in industry, construction, agriculture for heat shield and sound insulation [1]. Their properties are described by the following characteristics: the bulk density $(\frac{T}{m^s})$; the particle size (mm); the friction coefficients (internal, with steel, with wood, with rubber); the angle of inclination of the slope (degree). They describe operation properties of bulk materials, their use for various technical purposes and in different industry sectors [2-3]. However, exactly acoustic properties – the speed of propagation and attenuation of sound, must be defined for determination of sound insulation properties. Such instruments and methods of measurement of acoustic characteristics of bulk materials are missing currently. This is due to physical difficulty sounding samples due to the very large ultrasound attenuation in them. Instrument sounding method implementation presupposes the existence of various structural elements. The appearance of any mechanical contact between them leads to acoustic noise, the amplitude of which significantly exceeds the levels of acoustic signals after passing through a layer of bulk material.

2. Expanded Perlite

Perlite is a highly efficient heat-and sound ecologically clean material, reliable servant in the temperature range from -200° to +900° C. Swelling perlite is carried out in furnaces by thermal shock at 900-1100° c. the presence of constitutional water attached Perlite's ability to Slough when heated. Bound water, evaporating, creates countless tiny bubbles in the softened mass. The breed is divided into spherical grain with an increase in volume 4÷20 times and porosity of up to 70÷90%. Change thermal expansion mode allows to produce various materials ranging from fractional composition of Perlite powder (less than 0.14 mm) to perlite crushed stone (10-20 mm). Depending on the size of grains of wood flour are distinguished (0.1 mm), perlite (0.16 mm), agroperlite (1-5 mm). Perlite sand is widely used in insulation fillings walls, floors and ceilings. Perlite bulk density ranges from 75 to 200 kg/m^3 use of Perlite reduces the thickness of the thermal insulation filling compared to the widely used ceramic concretes in Russia in 2-3 times. Compared to other thermal insulation materials, ecologically pure sand expanded perlite, incombustible, does not age, it does not live pests. Perlite is also used in glass, metal, chemical industries, agriculture (agroperlite).

Heat and sound insulating properties of expanded perlite defines its structure. For control of technological process, as well as to obtain different fractions of material structure analysis becomes necessary at different stages of production.

To get an idea of the internal structure of the material using x-ray computed tomography was used to capture the image slicer (tomogramma) perlite powder (Figure 1A) and 3-D image volume Ø5 x 5 mm (Figure 1B). Picture shows that expanded perlite is a combination of particles and their shape is close to areas of different diameters are available "stick together" sphere; also of different diameters are enclosed fields. A significant portion of the volume of matter takes the air. The term "structure" describes characteristics: diameter range of spheres; average diameter volume areas; thickness of individual areas; the relative content of material and air.

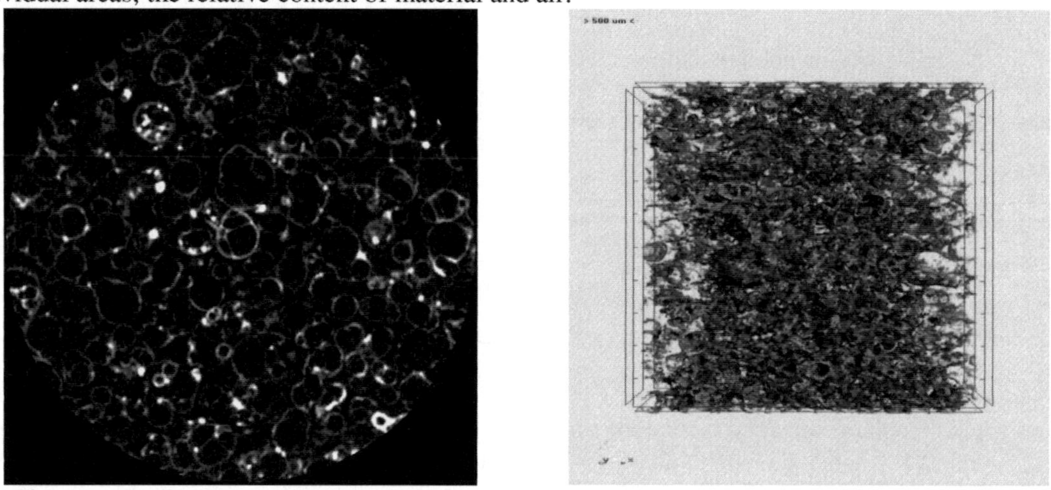

Figure 1. Tomograms, where a) tomogramma sample of expanded perlite; b) 3D-image element volume of expanded perlite.

In the work we investigated the possibility of application of ultrasound to monitor the structure of expanded perlite. The analysis showed that the information about the characteristics of the structure of the material can be obtained through measurement of sound velocity in it and its attenuation. However, measuring the speed and attenuation of sound in bulk materials using available modern equipment and measurement techniques impossible. This is due, primarily, with a very large attenuation in any bulk material. Secondly, it is impossible to carry out acoustic emitter and receiver at the same time contact with the powder. Or input or reception must attend an air gap, which excludes mechanical contact between the individual elements of design.

3. Installation

For measurement of velocity and attenuation of sound installation was developed for measuring the acoustic characteristics of granular materials. Installation work is based on the measuring time and attenuation of the wave in a layer of material known thickness, when passing the sound further, through well-known layer of air. Speed and sound attenuation in bulk material are determined by comparison with speed and attenuation of sound in air. Structural diagram of an apparatus for measuring speed and ultrasound attenuation in bulk material is shown in Figure 2.

The installation consists of the following units: a piezoelectric emitter (1), piezoelectric receiver (2) with integrated preamp. Probe excited by pulse generator output-device UK-14PM (3) and creates an acoustic pulse that passes through the material (h) and (H) and arrives at the receiver (2), passing through the zone control $(L)_0 = (h) + (H)$.

At the output of the receiver Crystal (2) there is an electrical impulse that, amplifying preamplifier fed to the input Attenuator (4). The Attenuator (4) is used to measure the amplitude of the past through the control zone of the acoustic pulse and adjust the signal level at the input connector of the device of the UK-14PM (3) material is controlled in the cell height (h). The ditch is installed on the radiator. PROBES through the contact layer of oil.

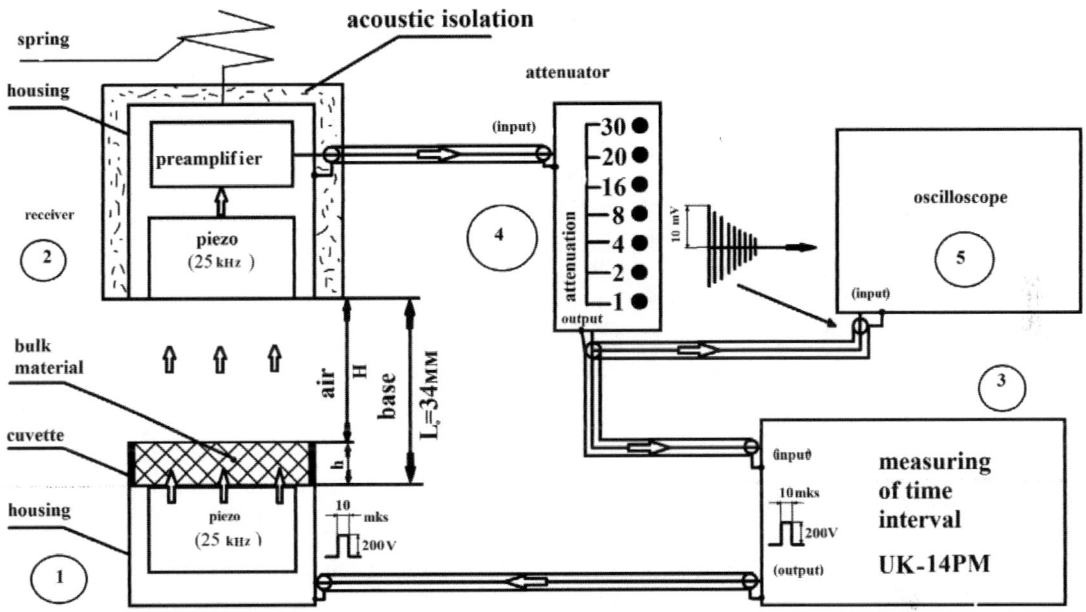

Figure 2. Block diagram of the Setup for the measurement of velocity and attenuation of sound in bulk materials.

Receiving the converter cannot be installed on the top surface of the material, because controlled amplitude spurious signal passing through the walls of the cells is much higher than the amplitude of the signal passing through the measured material. The only possibility of measurement is the measurement of acoustic signals from the top surface of the controlled material thickness (h) and passed through the layer of air H. While receiving the Converter requires a very effective sound insulation from all elements of the installation. The high attenuation of sound in bulk material and in the air receiving signal from preamplifier gain not less than 10^3. To measure time intervals used appliance UK-14PM. It is designed to work with converters from 20 kHz to 1 MHz and is designed for measuring the propagation time of longitudinal ultrasonic oscillations shadow method. The device provides the measuring range the propagation time of 20 μs up to 8800 μs and time measurement accuracy 0.01 t.

3. Sound velocity measurement.

The process of measuring the speed of sound consists of the following steps:

- Calibration time interval meter UK-14PM;

To calibrate the meter is used with State standard sample SO-2 with time passing longitudinal wave 10 μs. The sample is installed between the emitting and receiving transducers and with potentiometer "mouth. 0 "On the display of the device is set reading" 10.0 μs.

- Measurement of the speed of sound in air;

Measurement of the time passing (t_{air}) ultrasonic pulse through air gap (L_0) without measurable material. Of values (t_{air}) and (L_0) is speed of sound in air is defined as

$$C_{air} = L_0 / t_{air} \quad (1)$$

• Is poured into the cavity height (h) bulk material. To move full time is measured **t**. At a known velocity value. sound in air full time is

$$t = t_{mat} + H/C_{air} \quad (2)$$

• Determined by the moving sound through a layer of granular material (h) and calculated the speed of sound in it.

$$t_{mat} = H/ C_{air} \quad \text{and} \quad C_{mat} = h/(t - H/ C_{air}). \quad (3)$$

Table 1 shows the average results of measurements of the speed of sound in expanded perlite for a fraction of 0.1-0.2 mm (filtroperlite).

Table 1. Speed of Sound in Perlite.

L_0, [mm]	h, [mm]	H, [mm]	C_{air}, [m/s]	T, [μs]	t_{mat}, [μs]	C_{mat}, [m/s]	Error, Δc, [m/s]	Error Δt, [μs]
34	10	24	322	114	39.7	252	± 5	± 0.3

As expected, the speed of sound in any bulk material must be less than the speed of sound in air, since the passage of ultrasound through the loose material describes the movement of sound through air gaps between grains resulting trajectory. US produces a curved, its length is greater than the geometric thickness of material (h). Through the grain themselves sound practically does not pass, because grains have solid sides, on which there is a scattering of sound. Evidently, the speed of sound propagation in bulk material depends on its structural characteristics, in particular medium-sized grains. Studies on the different fractions are represented in Figure 3. They show that with the increase of the average grain size of ultrasound velocity increases, since less effective length of trajectory of sound in the material [4-7].

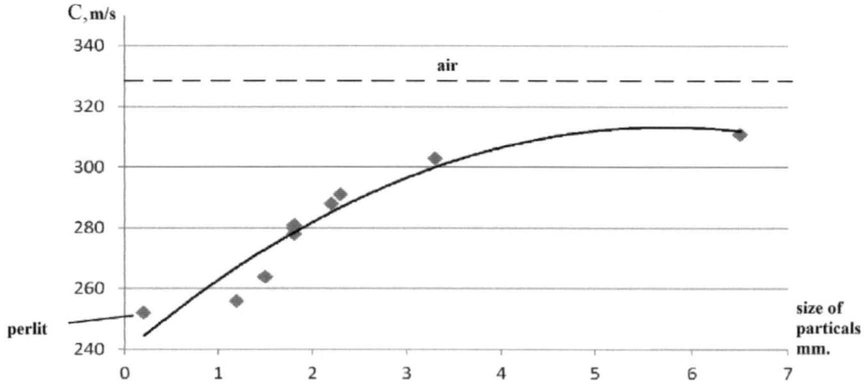

Figure 3. Dependence of the velocity of sound in perlite to different fractions.

Size fractions of 0.1- 0.2 mm corresponds to the speed of sound, 252 m/s. Thus, by measuring the velocity of sound, you can control the average grain size, i.e. the fraction of Perlite. Used information allows you to control (regulate) thermal process technology "heaving".

For the measurement of sound attenuation in the installation Figure 2 used attenuator (4). The measurement process is based on the fact that the full sound attenuation measured in dB (N) consists of attenuation in the material N_{mat} and attenuation in the air N_{air} [4, 8]. I.e. you can write

$$N=\delta_{mat}*h+\delta_{air}*H, \quad (4)$$

where δ_{mat} is an attenuation in the material (dB/mm); δ_{air} is attenuation in the air (dB/mm); (h) and (H) are in millimetres.

4. Measurement Attenuation

Attenuation measurement technique consists of the following steps:

- Determination of the coefficient of attenuation in the air;

Instead of bulk material installed steel sample thickness (h).

The used frequency 25 kHz attenuation in steel is negligible [9, 10] sound will take place with the attenuation layer of air (H) and the attenuation will be $N(H_{air})$.

Therefore, if you change the path of the sound L_0–H=h, changing the attenuation will be

$$\Delta N_{air}=N(L_0)_{air}-N(H)_{air} \quad (dB)$$

and damping factor in air determined

$$\delta_{air=} \Delta N_{air}/(L_0-H) \quad (dB/mm). \quad (5)$$

- Determination of the coefficient of attenuation in the material.

For this loose material is poured into the cavity height (h) and measured total attenuation (N) (dB). Attenuation in the material will be defined as

$$\delta_{mat}=(N-\delta_{air}*H)/h. \quad (6)$$

Measured average attenuation coefficient US perlite 25 kHz frequency is 19 dB/cm or in relative units 2.185 Np, that far exceeds the damping factor at the same frequency in the air 5.42 dB/cm (0.624 Np) [11-13].

5. Summary

The developed method and installation of measure speed of propagation and attenuation of sound in bulk materials. On the example of perlite it is shown, that for the purpose of monitoring its structure and determine the fraction can be used ultrasound velocity through the material and its attenuation. It has been shown that with increasing grain size of the material, the speed of sound increases, but it cannot exceed the speed of sound in air. A method for measuring sound velocity and attenuation in perlite (and other loose materials) is. Analogues of such installations do not. Installation and measurement technology of the speed of sound and attenuation can be used by developers of heat and noise insulating materials to improve their technologies.

References

[1] Dzhenike E V 1968 *Warehousing and bulk materials release* (Moscow: Mir) (in Russian)
[2] Komar A G 1988 Building materials and articles (Moscow: Lebedevka) (in Russian)
[3] Zhukov A et al 1960 *Circulite* (Kiev: Gosstrojizdat) (in Russian)
[4] Krasilnikov V J 1960 *Sound and ultrasonic waves in the air, in the water and solids* (Moscow: Fizmatlit) (in Russian)
[5] Sun J *et al* 2016 *RSC Adv.* **6** 29220–29228 DOI: 10.1039/C6RA02933G
[6] Zhong Y *et al* 2017 *MATEC Web of Conferences* **102** 01038 DOI: 10.1051/matecconf/201710201038
[7] Rogachev A *et al* 2013 *RSC Adv.* **3** 11226–11233
[8] Kapranov B I *and* Korotkov M M 2008 *Acoustic methods of monitoring and Diagnostics* (Tomsk: TPU Publishing House) (in Russian)
[9] Truell R *et al* 1969 *Ultrasonic methods in Solid State Physics* (New York: Academic Press)
[10] Plotnikova I *et al* 2016 *IOP Conf. Ser.: Mater. Sci. Eng.* **132** DOI: 10.1088/1757-899X/132/1/012023
[11] Kolesnikov A E 1970 *Ultrasonic measurements* (Moscow: Izdatel'stvo standartov) (in Russian)
[12] Budenkov G A *et al* 1998 *Rus. J. of Nondestructive Testing* **34(9)** 631-635 ISSN: 10618309
[13] Moskovenko I B 2002 *V Mire Nerazrushayushchego Kontrolya* **2** 26-28

IOP Conf. Series: Materials Science and Engineering **189** (2017) 012033 doi:10.1088/1757-899X/189/1/012033

Focused Ion Beam Methods for Research and Control of HEMT Fabrication

E Ph Pevtsov[1], A V Bespalov[2], T A Demenkova[3], and P A Luchnikov[4]

[1]Director of Design Center of Integrated Circuits, Nanoelectronics Devices and Microsystems, Moscow Technological University (MIREA), Moscow, Russia
[2]Director of Nanocentre, Moscow Technological University (MIREA), Moscow, Russia
[3]Associate Professor, Moscow Technological University (MIREA), Moscow, Russia
[4]Senior Researcher, Moscow Technological University (MIREA), Moscow, Russia

E-mail: pevtsov@mirea.ru

Abstract. The combination of ion-beam spraying and raster electronic microscopy allows to receive images of sections of defects of the growth nature origin in epitaxial films on GaN basis with nanodimensional permission, to carry out their analysis and classification irrespective of conditions of receiving. Results of application of the specified methods for the analysis of technological operations when forming the microwave transistors are considered: formations of locks, receiving of holes and drawing of contacts.

1. Introduction

Electronic and microscopic methods are the powerful instrument of selective control of parameters of heteroepitaxial structures and substrates, and are especially urgent for improvement and optimization of the modes of engineering procedures of devices production. An urgent task is the development of techniques of incoming and technological control of heterostructures AlGaN/GaN/SiC [1]. Defects of substrates, interstices, influence of impurity at formation operations and other harmful factors introduced by technological operations interfere with receiving of the quality characteristics HEMT. Elimination of the specified shortcomings causes need of creation of techniques of the researches allowing to observe cross sections and to determine conditions of working structures in the chosen place. Combination of raster electronic microscopy of high resolution and preparation of objects of focused ion beam allows to carry out the preparation considering features of the researched object.

2. Aims and Definition of Task

The developed techniques and the equipment are intended for works on the control of quality of the substrates of silicon carbide which are used for production of heterostructures AlGaN/GaN/SiC, and also the heterostructures which are applied to creation of the microwave GaN HEMT devices. In this work the description of solutions of the following tasks is stated: 1) selective input control of the substrates intended for devices production of heterostructures on the GaN basis; 2) development of techniques of technological postoperative control of plates with heterostructures for optimization of technological procedures of an electronic lithograph, etching and metallization.

IOP Conf. Series: Materials Science and Engineering **189** (2017) 012033 doi:10.1088/1757-899X/189/1/012033

3. Technique of Researches

In this research with the equipment Helios NanoLab the following technological operations are developed and applied: the local ion-beam etching based on superprecision ionic dispersion under the influence of an focused ion beam at the high accelerating voltage; the ion-stimulated sedimentation based on local ionic activation of process at introduction to a zone of processing of chemically active gas, a source of the besieged material. The resolution of technological operations and the resolution in the microscopic mode of observation in secondary electrons realized by means of an ionic bunch of preparation made 7 nanometers and 0,4 nanometers respectively. At the same time the focused ion beam was applied to creation of cross sections of a sample perpendicular or at an angle to his surface. The analysis of cross section of a sample has allowed to distinguish borders between the materials received at different technological stages, local heterogeneity and foreign inclusions. Observation of features of structure in the depth of samples can be made not only in static, but also in the dynamic mode at ionic etching and simultaneous visual control by an electron beam.

Process of creation of cross sections included the following main stages: 1) obtaining the orthogonal image of an object (cut) for the choice of the site of preparation; 2) providing a good drain of a charge for obtaining qualitative images; 3) the rough etching deleting the main part of material before the studied object; 4) the polishing etching revealing thin structure of an object of a research; obtaining image of an object. As a result the typical cut of the studied structures represented a rectangular well up to several microns in depth, one of walls of which is polished.

Current research shows that in order to obtain high quality cross-sectional images of the grown defects in epitaxial GaN layers with the use of FIB, the removal of effects caused by the violation of the material stoichiometry is needed, because preparation and polishing by ion beam comes with implant of GaN ions related to the layer material. Resulting not-stoichiometry compound Ga1+xN was decomposing with excreting of metallic Ga in the form of drops. To remove this effect the reduction of energy and ion beam current was used with local deposition of technological platinum layer about 0.5 micron thick on the area of the defect. The platinum layer prevents ion implantation of Ga which eliminates metallic Ga drops formation. That allows increasing the contrast of image and reducing the effects of the charging of the dielectric surface. The deposition of the metal layer for the drain charge at determined surface areas was made by the source of metal-organic platinum compound.

3. The Obtained Results and Their Discussion

3.1. The analysis of initial structures GaN

Deep etching by an ion-beam method was applied to the layer-by-layer analysis of structural features of samples GaN. The modes of this process have been picked up so that they were compatible to standard operations of formation of structures of microelectronics. As a result it is received that layer-by-layer ion-beam drain allows to reveal the latent defects like emptiness and punctures which have partially geometrical facet (a hexagonal form of a pole of etching and a pyramid of growth). Existence of the latent defects like punctures is the reason of impact of implantation on depth at 5-15 times the exceeding length of a run of argon ions in gallium nitride (according to data of atomic and power microscopy and a photoluminescence) as the speed of their dispersion is 2-4 times higher than the speed of dispersion of an initial surface. The latent defects are the main channels for leakages of current after metallization. The combination of ion-beam dispersion with the FIB and SEM methods allows to receive images of sections of defects of the growth nature of an origin in epitaxial films on the GaN basis with the nanodimensional permission, to carry out their analysis and classification, irrespective of receiving conditions. Table 1 generalizes the obtained data.

During the testing of defect identification method it was revealed [2] that growth defects in layers with density higher than $10^6 cm^{-2}$ are caused by the lack of epitaxial agreed substrates. They adversely affect the operating characteristics of the device structures manufactured by planar technology. Electron microscopy techniques allowed detection of large group of defects in the form of hexagonal pyramids, positioned with their base to the surface, with typical dimensions of the base up to 200 nm,

IOP Conf. Series: Materials Science and Engineering **189** (2017) 012033 doi:10.1088/1757-899X/189/1/012033

penetrating even in the substrate. GaN on Si films contain a large number of longitudinal cracks along the direction (001). Physical proximity of cracks (less than 1 μm) causes the separation of layer material from the substrate.

Table 1. Classification of growth defects in GaN based on their distribution by size and depth.

Defect type	Defect description	Defects density	Electron microscope image
I	Small defects (less than 10 nm) in near-surface epitaxial p-GaN layers	$10^8 \div 10^9$ cm^{-2}	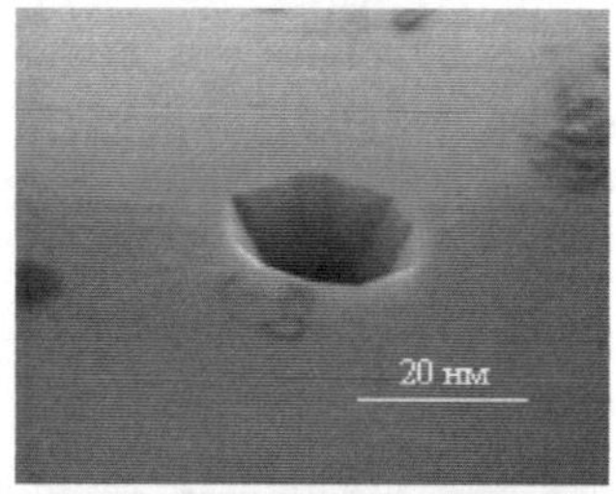
II	Prolonged defects (up to 200 nm) through epitaxial p-GaN layer	$10^7 \div 10^8$ cm^{-2}	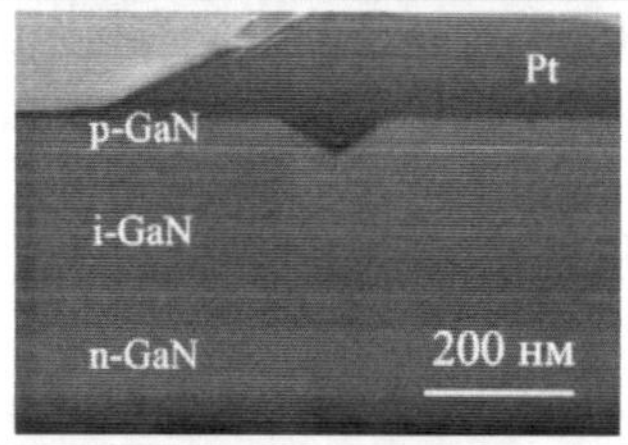
III	Defects going through multilayer epitaxial GaN structure (up to 3,5 μm)	$10^6 \div 10^7$ cm^{-2}	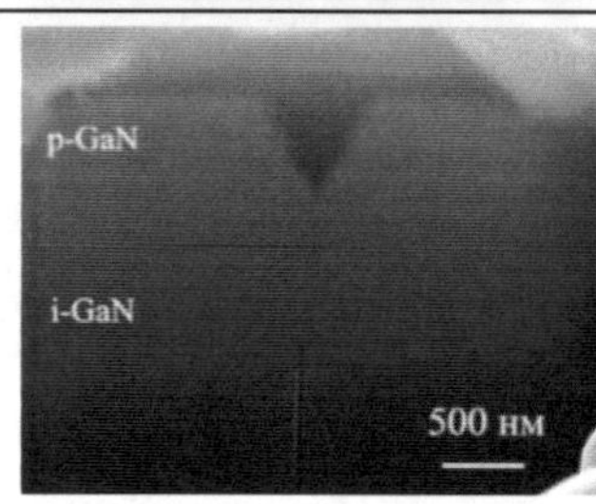

3.2. Control of T-shaped Lock Formation Processes

In recent years, speed increase of devices based on wideband gap semiconductor materials is achieved mostly with the decrease of gate length. That also allows increasing transistor transconductance and decreasing noise figure [3].Basically microwave field transistors with Shottky barriers utilize so-called T-shaped gate. Traditionally electron-beam lithography with three-layered resist system PMMA/(PMMA&PMAA, copolymer)/PMAA is used in creation of T-shaped gates. Since that technique is based on the difference in exposure speed of copolymer and PMMA, during the formation of T-shaped gate the size of its pin will be strongly influenced by the exposure time, copolymer thickness and it chemical composition [4–8].

During testing of T-shaped gate formation process, samples of HEMT patterned gate structures with multilayered resists on different substrates (silicon, silicon carbide, sapphire), and with gate metallization, were investigated as test samples. FIB was used to obtain resist exposure profiles, to determine thickness and homogeneity of gate metallization and to reveal the resulting defects. Typical profile of T-shaped gate with three-layered resist system is shown in Figure 1.

The main results of researches of structures with the T-shaped lock have shown that defect when forming T-shaped locks is caused by the following major factors: 1) non-uniformity of resist exposure; 2) effects of contamination of edges; 3) defects of metallization.

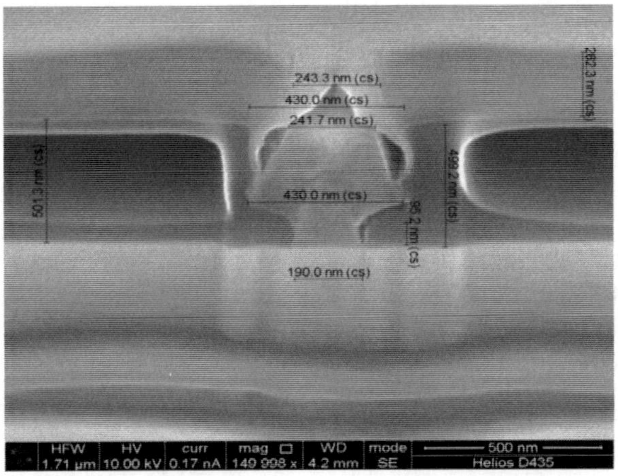

Figure 1. Cross-section of T-shaped gate structure.

3.3. Control over Hole Formation Processes

In transistors on the basis of heterostructures AlGaN/GaN made on silicon carbide apply through metallization that allows to take away effectively heat and reduces parasitic inductance. For this purpose apply jet and ion etching, in particular, etching in the inductive and connected plasma. The technique developed on the basis FIB have allowed to estimate depth of the etched openings, a tilt angle of walls and their morphology, and also to define the defects which are formed in the course of etching. Tests were performed on samples on semi-insulated 6H-SiC substrates 50 mm in diameter and thickness of 100 µm with back-side epitaxial GaN/AlGaN layers 2 µm thick. Typical cross section of SiC via hole with etching stopped at GaN/AlGaN layer is shown in Figure 2.

Figure 2. Cross-section of 6H-SiC surface after the etching of non-through holes stopped at GaN/AlGaN layer.

Hole depth is 100 µm, with ribbed side-walls. Ridges with different height are visible at the bottom, which are presumably pillars, appeared as a result of micromasking effects on micropipes and screw dislocations. Maximum height of the defects is 25 µm. Except that, the surface of gallium nitride could be observed, which characterizes by slow etching in fluorine based plasma.Two types of post-etching SiC surface defects were observed during the research, distinguishing by different distribution on the surface. The defects were in a form of round holes with transverse dimensions up to 50 µm. Some defects have distinct hexagonal form. Similar defects were observed in [9]. At the same time on the bottom of not through holes defects in the form of deepenings of irregular shape are observed. In the center of some defects there are small single ledges.

3.4. Control of Ohmic Contacts

Creation of ohmic contacts meeting the requirements mentioned above for GaN based devices technology is still the subject of research [10]. Main requirements for ohmic contacts are smooth surface morphology, sharp (smooth) edges of contact pads, thermal stability and current-voltage characteristic linearity. Ti/Al/Ni/Au is frequently used as of ohmic contact to GaN [11]. Flaw of such system is high degree of surface roughness due to strong mixing of Au and Al. In this research, samples for the testing of methods for control of AlGaN/GaN/SiC device defects had ohmic contacts Ti/Al/Ni/Au and Mo/Al/Mo/Au. Typical thickness of contact structures are: Ti(20nm)/Al(100nm)/Ni(100nm)/Au(200nm). It is shown that areas of ohmic contacts containing Ti have protuberance-shaped defects, typical for melt of AlNiAu compounds, and the interface with titan sub-layer is not homogeneous and contains foreign inclusions in the form of separate cavities, whose size can be up to 30-50 nm. Microscope images of annealed Mo/Al/Mo/Au and Ti/Al/Ni/Au ohmic contact surface are shown in Figure 3.

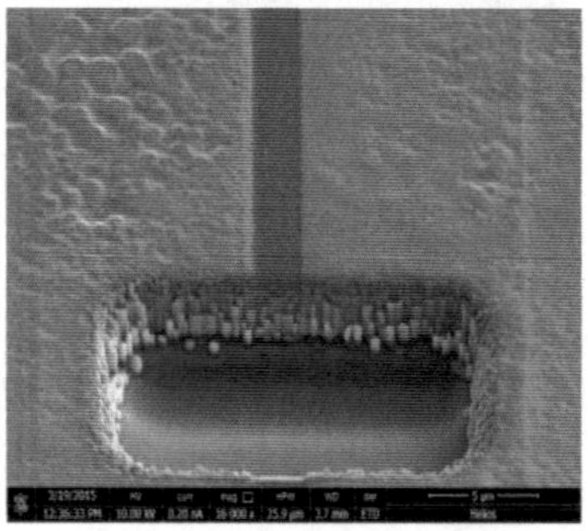

a)

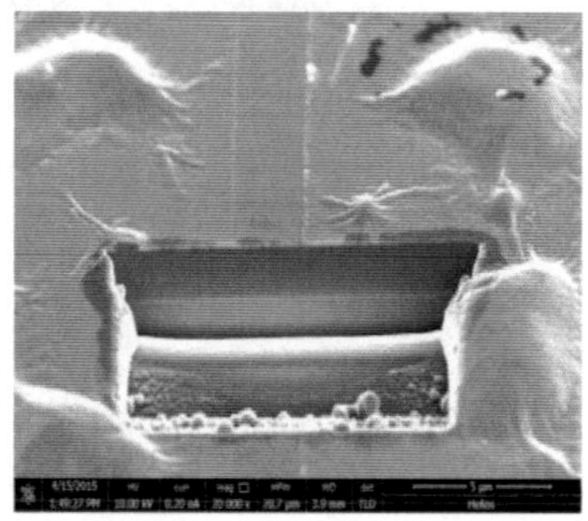

b)

Figure3. Microscope images of Mo/Al/Mo/Au (a) and Ti/Al/Ni/Au ohmic contact surface (b).

On typical microscopic images of cross-section of the ohmic contact Mo/Al/Mo/Au the boundary of layers GaN and AlGaN is well noticeable and layers of heterostructure and metallization are well perceptible. Metallization and AlGaN layer thicknesses are in a good agreement with data obtained during deposition and growth, respectively. Melting of metals was not observed, compared with conventional Ti/Al/Ni/Au [12].

4. Conclusion

The developed techniques of research of initial plates defects and technological control of formation of devices with heterostructures were used for tasks of optimization of technological processes of electronic lithograph, etching and metallization.

Authors consider that in this work the following provisions and results are new:

The technique of researches of structures with GaN combining raster electronic microscopy of high resolution and nanodimensional local preparation by the focused ion beam is developed.

Data on standard defects of growth in epitaxial layers GaN, their distribution on a surface and volume are obtained. Various defects of the substrate 6H-SiC are revealed and classified.

The technique of quality control of drawing a rezist for formation of T-shaped locks is offered.

Control methods of technological operations for receiving contact holes in SiC to a working layer GaN are developed and approved.

Recommendations about optimization of the modes of technological operations for creation of ohmic contacts Ti/Al/Ni/Au and Mo/Al/Mo/Au with the acceptable morphology of metallization and rather low specific resistance are formulated.

Acknowledgments

Research was supported by the Ministry of Education and Science of the Russian Federation (No 8.5098).

Authors would like to express their gratitude to the specialists of "PULSAR" S&PE" JSC A.A. Dorofeev and F.M. Baryshnikov for their assistance in preparation of article materials and valuable comments.

References

[1] Gruzdov V V *et al* 2016 *Control of new technologies in solid-state microwave electronics* (Moscow: Tekhnosfera) (in Russian)
[2] Bespalov A V *et al* 2016 *Proc. of Int. Conf. Microelectronics'15* 373–382 (in Russian)
[3] Yoon H S *et al* 2015 *Proc. of Int. Conf. CS ManTech'15*
[4] Luchnikov P A *et al* 2016 *IOP Conf. Series: Materials Science and Engineering* **110(1)** 012081 DOI: 10.1088/1757-899X/110/1/012081
[5] Pevtsov E Ph *et al* 2017 *IOP Conf. Ser.: Mater. Sci. Eng.* **168(1)** 012095 DOI:10.1088/1757-899X/168/1/012095
[6] Aleutdinov A D *et al* 2015 *IOP Conf. Series: Materials Science and Engineering* **81(1)** 012069 DOI: 10.1088/1757-899X/81/1/012069
[7] Luchnikov P A 2015 *Russian Physics Journal* **57(9)** 1239–1245 DOI: 10.1007/s11182-015-0369-0
[8] Piliptsou D G *et al* 2017 *IOP Conf. Ser.: Mater. Sci. Eng.* **168(1)** 012103 DOI:10.1088/1757-899X/168/1/012103
[9] Guy O J *et al* 2008 *Applied Surface Science* **254** 8098–8105
[10] Sasakura A *et al* 2015 *Proc. of Int. Conf. IMFEDK'15* 42–43
[11] Jung S M *et al* 2015 *Semiconductor Science and Technology* **30(7)** 075012
[12] Shcherbachev K D *et al* 2015 *Electronic Engineering. Series 2* **4(238)** 38–48 (in Russian)

IOP Conf. Series: Materials Science and Engineering **189** (2017) 012034 doi:10.1088/1757-899X/189/1/012034

Phytoremediation of Excavation in Eastern Siberia

S S Timofeeva[1], and T M Zhmurova[2]

[1]Professor, Irkutsk National Research Technical University, Irkutsk, Russia
[2]Graduate student, Irkutsk National Research Technical University, Irkutsk, Russia

E-mail: timofeeva@istu.edu

Abstract. The paper deals with prospects and applicability of phytoremediation technology in the conditions of Eastern Siberia as applied to gypsum stone extraction of Zalarinsky deposit in Irkutsk region. It analyses dust intensity for different technological processes of the deposit and evaluates the dust loading on the workers of the deposit. The selection of plantings that have potential to neutralize the dust loading produced by the plot "Severniy" is carried out.

1. Introduction

At the present time increasingly greater attention is paid to the creation and introduction of the bioremediation technology represented a battery of methods of a purification of water, soil and atmosphere applying such biological objects as mushrooms, plants, worms, insects and other organisms [1].

There are three fundamental approaches to the soil bioremediation that operate by using microorganisms:

- biostimulation – the development of indigenous microflora in the polluted areas;
- bioodorange – the introduction to soil biological preparations and microorganisms which contribute to degrade the pollutant;
- phytostimulation – the stimulation of the microorganisms' development with plants.

Microorganisms are important in the degradation of the pollutants. Plants are biofilters that create the environment, contribute to the soil loosening and provide oxygen. This technology begins to be applied in mining particularly the methods of phytoremediation are widely used.

Phytoremediation is an energy-efficient aesthetic way to restore an area with low or medium level of pollution. Plants can accumulate heavy metals by gathering them in the leaves from polluted soils; sorb on the leaves' surface a fine-dispersed dust emitted during industrial processes; strengthen the surface of the depleted pits; prevent soil erosion, reduce noise, etc.

In [2] the review scheme of phytoremediation is shown (Figure 1). Filtration function of the green plantations consists of the mechanical retarding of the dust and chemical compounds from the natural environment and thereby the air purifies from pollutants.

The methods of phytoremediation are devising and applying actively in Bulgaria, the USA, the UK, Spain, Canada, China, Mexico, New Zealand and several other countries [3–4].

For example, it was calculated that in Beijing (China) 772 tons of dust are removed by trees annually in the city core [5]. In Shanghai (China) the decreasing of the dust concentrations by 9.1% was observed. MacDonald and others [6] have demonstrated that planting trees in the West Midlands (the UK) from 3.7% to 54% of the accessible land can involve the decreasing of the dust concentrations in the air by 26%; it means that the removal of dust is about 200 tons per year. In

Chicago the trees that occupy 11% of the city area remove approximately 234 tons of dust per year [7], in the USA in general trees and bushes of urban areas absorb about 215000 tons of dust per year.

In Russia the introduction of the nature conservation technology has been impeded extensively by a necessity to search the plants peculiar to our soil and climatic conditions. By now the facts about abilities of single species of wild plants as well as cultivated tree and shrubbery plants along with herbaceous plants are accumulated and a wide range of plants that are suitable for soils' phytoremediation of different type pollutants is revealed.

The purpose of the research is to evaluate prospects and applicability of phytoremediation technology in the conditions of Eastern Siberia as applied to gypsum stone extraction of Zalarinsky deposit in Irkutsk region.

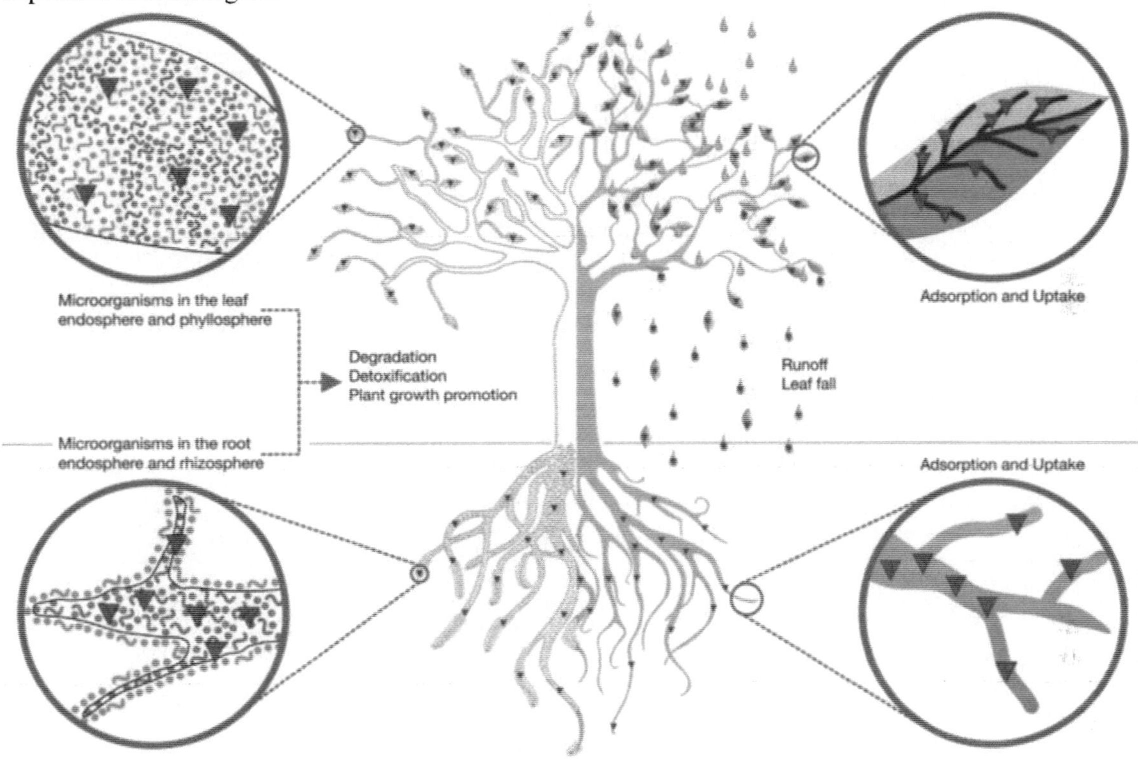

Figure 1. The review scheme of air pollution phyto-treatment.

On the basis of the production activities analysis of domestic enterprises of mining and processing of non-metallic materials we have come to a conclusion that enterprises have exactly the same technological processes such as mining, drilling-and-blasting and crushing. These technological processes are accompanied by dust emission, noise and vibration as well as a significant impact on the environment. Therefore, for the analysis and calculation Zalarinsky gypsum deposit was chosen as it is one of the leading deposit for the gypsum stone excavation. The deposit is the youngest and most promising in Eastern Siberia. Reserves of gypsum will be enough for at least 50 years. Zalarinsky gypsum deposit is exploited by the German company KNAUF.

The plot "Severniy" of Zalarinsky gypsum deposit is located in Nukutskiy district of Irkutsk region 20 km from Zalari railway station. The purpose of the enterprise is a preparation, mining and extraction of gypsum stone of Zalarinsky gypsum deposit by open-cut and split method [8].

The analysis of the dust intensity for different technological processes is carried out and the applicability of phytoremediation technology to neutralize the negative impact of the enterprise on the environment.

IOP Conf. Series: Materials Science and Engineering **189** (2017) 012034 doi:10.1088/1757-899X/189/1/012034

2. Results and discussion

The dust content in the air was evaluated by the gravimetric method by means of glasses that were displayed during the day both close to operation zones of manufacturing equipment and the industrial area outside the plot. It has been found experimentally that the maximum dusting is fixed during crushing, loading and transportation of mass (Table 1). About 62 % of all emissions at the open pit is inorganic dust with silicon dioxide.

Table 1. The dust intensity for different technological processes of extraction of gypsum stone on plot "Severniy" of Zalarinsky deposit.

Technological process	Dust intensity, gps	Mass of emitted dust, t/a
Drilling-and-blasting operations	0.0002	0.0056293
Crushing	0.062	1.95718
Loading operations	0.056	1.74589
Transportation	0.031	0.97154
The area outside the plot	0.35	11.07318

Outside the plot the dust content in the air ranges from 15 to 30 mg/m^3. The emission of the fine-dispersed inorganic dust with SiO^2 into the atmosphere generated in the process of mining and crushing of gypsum stone amounts to more than 16 tons per year.

Table 2 shows the results of the dust load calculation of the workers who work for the crushing plot of the gypsum stone of Zalarinsky deposit.

Table 2. The results of the allowed active time of the workers who work for the crushing plot of the gypsum stone of Zalarinsky deposit.

The name of a profession	Dust concentration, [mg/m^3]	MPC, [mg/m^3]	Control dust loading	Dust loading	Allowed active time, years
An equipment operator of mining and crushing	4.1	2	124 000	254 200	12.2
The engineer of crushing station	7.2	2	124 000	446 400	6.5
The engineer of belt	4.1	2	124 000	254 200	12.2

It has been established that the active time without showing any pathological changes in the lungs for described professions is at least 6.5 years.

In order to reduce the dust from fugitive sources the selection of plantings that have potential to neutralize the dust loading produced by the plot "Severniy" is carried out on the basis of literature data analysis on the composition of trees and bushes that grown in the climatic conditions of the studying region [9–10].

Minimization of dust and air pollution by green plantations is in the result of the abeyance of the polluted stream movement rate in the woodland the dust fall and inhibition takes place on the surface of leaves, shafts and branches [11]. The rain or water filament washes away caked dust during irrigation of green plantations.

The studying plot "Severniy" of Zalarinsky gypsum deposit relates to the Balagansk-Nukutskiy district (Central Siberian forest-steppe province) where low power chernozem soils, carbonate and solonetzic, gray forest and caespitose-calcareous soils are distributed.

As a promising phytoremediators of this area it is possible to recommend the black poplar – a tree of the first rank reaching 30–35 m in height and 1–2 feet in shaft diameter. Black poplar is propagated by seeds, stool shoots and root stalks.

Black poplar is a rapid-growing tree which strikes root very well in different climatic zones. In addition to it black poplar stands up well prolonged inundation because the tree can absorb a lot of moisture from the soil.

For the vegetation period growing black poplar precipitates about 44 pounds of dust, white poplar - 53 kg; white willow and ash-leaved maple 34, 30 pounds respectively. One hectare of spruce forest precipitates 32 tons of dust per year [9].

Black poplar is the most widespread species of woody plants which is mainly used in biological recultivation, strengthening banks, river banks and for landscaping of bald populated areas. The reason is that it is very frost-resisting, rapid-growing and can withstand deep pruning along with it is wind-firm, environmentally flexible and steady to dust, smoke and gas. For the vegetation period one poplar dusts the atmosphere from 20–30 kg of dust or soot [9].

As phytoremediators of this site one can also plant the common lilac.

Lilac is a genus of plants of the olive family. These are low deciduous shrubs with entire-kind leaves [12]. This shrub is very resistant to drought, frost, industrial gas. Lilac emits volatile oils which have antibacterial properties. Young leaves of lilac hold five times more dust than leaves of birch. The best dedusting species of lilacs are Amur lilac, hairy and Japanese lilacs, but unfortunately, they are not introduced in the region.

As additional fitoremediation we have evaluated the possibility of using hydroseeding of herbs which can grow on sandy soils low in nutrients. Red clover and alfalfa can be the most unpretentious and perspective.

Red clover is a perennial undemanding herbaceous plant of the pea family in height up to 50 cm. It grows in meadows, along roadsides and also it can be found in the field, on slopes, along rivers [13].

Alfalfa is a perennial wild or sowing plant that belongs to the pea family. It has a thick strong stem, oval leaves and can grow in height up from 20 to 100 cm [14]. Its strong assemblage of rootlets spreads out large areas thus improving the structure and air permeability of the soil.

Alfalfa has useful properties due to its rich balanced composition and the presence of scarce amino acids and chemicals valuable to human health. As alfalfa assemblage of rootlets is from 15-20 meters in depth it pulls valuable nutrients (trace elements, minerals, etc.) from deep layers of the soil that are inaccessible to other plants.

The choice of plants arises from several reasons. First, these plants have a high productivity. Second, these plants form quickly a sod and thus protect the surface from wind and water erosion. Another reason is that in most cases these plants are not in need of soil fertility and most species withstand the lack of moisture in the soil. Finally, in the new economic conditions, the great advantages of these plants application are the availability of seed, simple sowing technology, minimal labour inputs.

3. Summary

Thus, the investigation of the dust intensity on plot "Severniy" of Zalarinsky gypsum deposit and the evaluation of prospects of phytoremediation technology suggested the application of this promising technology in the whole mining industry. At present the first experimental plantings of lilac and red clover have been conducted.

References

[1] Kireeva N *et al* 2011 *Proc. Samara Scientific Center of the Russian Academy of Sciences* **13(5-2)** (in Russian)

[2] Weyens N *et al* 2015 *Int. J. Mol. Sci.* **16(10)** 25576–25604 DOI: 10.3390/ijms161025576

[3] Jones R K *et al* 2004 *Environ Sci & Pollut Res* **11(4)** 266 DOI: 10.1007/BF02979635

[4] Kirk J *et al* 2005 *Environ. Pollut* **133** 455–465 DOI: 10.12691/ijebb-3-1-2

[5] Yang J *et al* 2005 *Urban Forestry & Urban Greening* **3(2)** 65–78 DOI: 10.1016/j.ufug.2004.09.001

[6] McDonald A G *et al* 2007 *Atmos. Environ* **41(38)** 8455–8467 DOI: 10.1016/j.atmosenv.2007.07.025
[7] Nowak D J 1994 *Chicago's Urban Forest Ecosystem: Results of the Chicago Urban Forest Climate Project, USDA Forest Service General Technical Report NE-186, Radnor, PA* 63–81 Retrieved from http://www.nrs.fs.fed.us/pubs/gtr/gtr_ne186.pdf
[8] Medvedeva S A and Zhmurova T M 2015 Proceedings of Irkutsk State Technical University 2(97) 75–83 (in Russian)
[9] Bakulin V T 2005 *Siberian Journal of Ecology* **12(4)** 563–571 (in Russian)
[10] Boyarkin V M *and* Boyarkin I V 2011 *Geography of Irkutsk Region* (Irkutsk: Sarma) (in Russian)
[11] Hjelmroos M 2000 *Proc. of 2nd European Symposium on Aerobiology* 703
[12] Rehder A *et al* 1928 *N.Y.: Mc Millan Comp* 777 – 783
[13] Tsao R et al J 2006 *J. Agric. Food Chem.* **54(16)** 5797–5805 DOI: 10.1021/jf0614589
[14] Anderson E 1953 *Biol. rev.* **28(3)** 280–307 DOI: 10.1111/j.1469-185X.1953.tb01379.x

IOP Conf. Series: Materials Science and Engineering **189** (2017) 012035 doi:10.1088/1757-899X/189/1/012035

High-Energy X-ray Radiation Registration Model

S P Osipov[1], O S Osipov[2], and K V Ignatenko[3]

[1]Leading Researcher, National Research Tomsk Polytechnic University, Tomsk, Russia
[2]Programmer, Company Solveig Multimedia, Tomsk, Russia
[3]Graduate Student, National Research Tomsk Polytechnic University, Tomsk, Russia

E-mail: osip1809@rambler.ru

Abstract. It is proved the necessity for carrying out computing experiments according to the mean value and the square of registered X-ray photons absorbed energy of in a scintillation detector. The offered imitation model of the transfer and registration of high-energy X-ray radiation in a sensing volume of CsI and $CdWO_4$ scintillation detectors is based on a Monte Carlo method. The model considers leakage of secondary photons and electrons. It is offered approaches to justification of adequacy of the developed model of a high-energy X-ray radiation registration.

1. Introduction

The present period is characterized by rapid systems development of digital radiography (DR), computed tomography (CT) and their applications [1, 2]. Non-destructive tests and customs monitoring of large-size objects by digital radiography and a computer tomography methods are impossible without sources of high-energy X-ray radiation. Betatrons and linacs of electrons are used as sources of X-ray photons with high maximal energy [3–6]. The main consumer characteristics of DR and CT systems are sensitivity, spatial resolution and efficiency. In the analyzed systems, rulers or matrixes of radiometric detectors, scintillation screens (panels) interfaced to a matrix of photo-detectors are applied as registrars of X-ray emission [7]. If elementary detectors of the linear and matrix registrars of radiation are isolate, and the radiating surface of photons sources has the small dimensions, then geometrical resolution of DR and CT systems is determined by detector sizes in the direction of perpendicular to a stream of quanta. Isolation is understood as absence or insignificance of optical and radiation interconnection of the next detectors. Optical signal interconnection from the next photo-detectors can be very significant for panel registrars. For this case, geometrical resolution of analyzed systems can be essentially less than the size of a contact surface of an elementary photo-detector. Effectiveness of filing of any kinds of registrars of X-ray radiation is determined by corresponding sensing volumes thickness in the direction of radiation propagation. Increase in thickness of scintillation screens conflicts with need of increase in geometrical resolution. Increasing the thickness of sensing volumes of elementary detectors is technologically difficult task for linear and matrix registrars of X-ray radiation. Thus, technical and technological correctness of using small effective sensing volumes of radiation and optical converters follows from the analysis of geometrical parameters of modern registrars of high-energy X-ray radiation. It is emphasized in the work [8] that high-energy X-ray photons leave a small part of energy when interacting with a scintillator that has small thickness or small cross sectional dimensions. Another part of photons energy is absorbed in the next detectors or transferred to external space. Secondary photons and electrons have high penetration ability therefore for primary interaction their filing not in the detector leads to deterioration in geometrical resolution of DR and CT systems. Measuring simple radiometric signals and appropriate noise levels remains one of the most important design problems of analyzed test systems. Calculation data of mean value and mean square dependences of registered photon absorbed energy by cylindrical scintillator radius based on various materials without regard to leakage of secondary electrons are given in the article [9]. An approach for approximating secondary electrons leakage is offered in the work [10]. The data obtained in the works [9, 10] are enough to estimate the mean value and the mean square of the registered photon absorbed energy for scintillation detectors which radius and thickness are more than 5 mm. For modern registrars of high-energy X-ray radiation, these data need to be corrected. The experimental solution of this problem is impossible because of a large number of parameters and high material cost. It means there is no alternatives to the modeling the X-ray radiation transfer by Monte Carlo method.

2. The Model of Photon Transfer Process in a Scintillator

The model of X-ray photons transfer in a substance by Monte Carlo method is based on the algorithms from the work [11] enhanced by blocks of secondary electron interactions with scintillator material.

2.1. Data for the Model

Assume infinitely thin X-ray radiation beam with continuous ionization energy spectrum falls along the axis on cylindrical scintillators with the radius r and thickness h. Scintillator material is CsI, $CdWO_4$. The classical Schiff's formula [12] is used for the description of high-energy spectra of X-ray radiation sources, various approximations are described in details in the article [13]. X-ray radiation maximal energies vary from 2 to 10 MeV. All main interaction effects of photons and substance considered in the model are as follows: the photo-effect, the coherent and incoherent scattering, the pair creation effect. The investigated random variable is the amount of absorbed energy left in a scintillator experienced interaction with a photon. Energy transfer of secondary photons and electrons is taken into account. The considered trajectories number N is set by a user.

2.2. Formation of the Preliminary Data

For increasing algorithm performance, the user database is formed at a preliminary stage on the base of 127-group library of data of interaction between gamma radiation and substance [14]. The data library [14] is a table of energy dependence of the total cross section of X-ray radiation interactions with substance $\sigma_{tot}(E_i)$ and interaction sections for photo-effect $\sigma_{foto}(E_i)$, Compton scattering $\sigma_C(E_i)$, Rayleigh scattering $\sigma_K(E_i)$ and pair creation $\sigma_{par}(E_i)$; the variable i varies from 1 to 127. The corresponding tables of the linear extinction coefficients $\mu_{tot}(E_i)$, $\mu_{foto}(E_i)$, $\mu_C(E_i)$, $\mu_K(E_i)$ and $\mu_{par}(E_i)$ are formed for the analyzed types of scintillators at the first stage. The user database consists of the spline-interpolation coefficients: a_i, b_i, c_i, d_i for the energy dependence $\mu_{tot}(E_i)$; $a_{foto\ i}$, $b_{foto\ i}$, $c_{foto\ i}$, $d_{foto\ i}$ for $\mu_{foto}(E_i)$; $a_{C\ i}$, $b_{C\ i}$, $c_{C\ i}$, $d_{C\ i}$ for $\mu_C(E_i)$; $a_{K\ i}$, $b_{K\ i}$, $c_{K\ i}$, $d_{K\ i}$ for $\mu_K(E_i)$; $a_{par\ i}$, $b_{par\ i}$, $c_{par\ i}$, $d_{par\ i}$ for $\mu_{par}(E_i)$. Furthermore, approximating functions which cumulative distribution functions of energy of E inverse to $F(E)$ considered as a random value are included in the user database. The table of the function $F(E)$ is calculated from the approximations [12, 13] of the numerical energy distribution $f(E, E_{max})$ for a X-ray radiation source with the maximal energy E_{max}

$$F(E) = \int_0^{E_{max}} f(E, E_{max})dE\,. \qquad (1)$$

Further, the table of the function $F^{-1}(E)$ is constructed from the table of the function $F(E)$. Let's notice that partitioning the energy range should not be more frequent than in the library [14]. Table spline-interpolation coefficients of the function $F^{-1}(E)$ are included to the user database: a_{fj}, b_{fj}, c_{fj}, d_{fj}.

2.3. Algorithm of the model of radiation transfer in a scintillator

The transfer model algorithm by Monte Carlo method comes down to the uniform draw of the trajectories with the same origin. The standard geometrical scheme of a draw of photon radiation transfer in a scintillator is represented in the figure 1 with the corresponding coordinate axis *XOY*. For convenience, let's consider polar coordinate axis with a lateral angle ω and a polar angle ψ. The red continuous line marks the trajectory of the primary photon. The red dash line marks the trajectory of the secondary photon. There are no such photons for the case of photo-effect, and there are exactly two photons with identical energy 0.511 MeV for the case of pair creation.
The secondary photon energy coincides with the primary photon energy for Rayleigh scattering but differs by the direction. The green marks a trajectory of a secondary electron on the Figure 1. For convenient, let's consider the auxiliary polar coordinate axis corresponded to the initial axis. The polar angles θ and θ' determine the difference between the primary photon and secondary particle directions (for a photon or for an electron). The corresponding azimuthal angles φ and φ' are equal to the angle between the vector's projection of the secondary particle movement to the plane *X'O'Y'* and *O'X* axis.

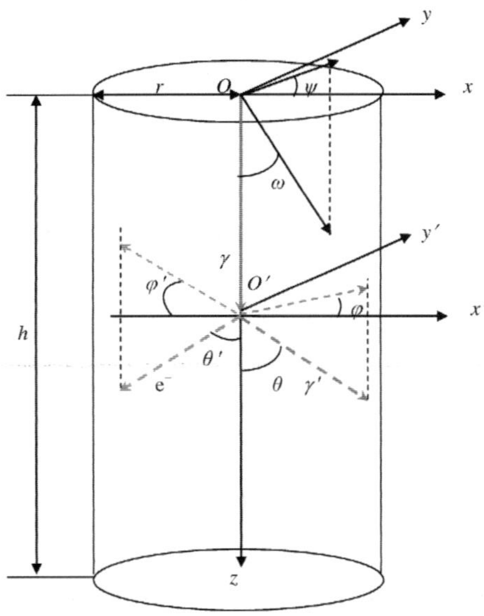

Figure 1. The geometrical scheme of a draw of photon radiation transfer in a scintillator

The surface of the cylindrical scintillator *S* is described by the expression

$$\mathbf{S} = \left\{ (x, y, z) : \begin{vmatrix} x^2 + y^2 = r^2,\ 0 < z < h, \\ x^2 + y^2 \le r^2,\ z = 0 \lor z = h \end{vmatrix} \right\}. \tag{2}$$

For increasing calculation efficiency, the corresponding averages were applied at each stage of photon interaction with scintillator material. Stages of the algorithm make a short list.

Zeroing of the photon counter i=0, the adders of the absorbed energy $\sum E_{ab}$ and the square of the absorbed energy $\sum E_{ab}^2$.

1. Cartesian and polar coordinates of photon penetration into a scintillator and coordinates a, b, c of a unit vector of the photon direction and weight factor w are as follows

$$\begin{cases} x = 0 \\ y = 0 \\ z = 0 \end{cases} \quad \begin{cases} \omega = 0 \\ \psi = 0 \end{cases} \quad \begin{cases} a = \sin\omega\cos\psi = 0 \\ b = \sin\omega\sin\psi = 0 \\ c = \cos\omega = 1 \end{cases} \quad w = 1. \tag{3}$$

2. Increasing the photon counter: $i=i+1$. Zeroing the interaction counter: $k=0$. Zeroing the local adder of the absorbed energy and the square of the absorbed energy of one registered photon: $E_{ab}=0$. A draw of photon energy by the method of inverse functions is given by the formula

$$E = F^{-1}(\xi) = \text{spline}(\mathbf{a}_f, \mathbf{b}_f, \mathbf{c}_f, \mathbf{d}_f, \xi), \tag{4}$$

where $\mathbf{a}_f$, $\mathbf{b}_f$, $\mathbf{c}_f$, $\mathbf{d}_f$ are columns of spline interpolation parameters for approximation of the function F^{-1}; ξ is a random number with uniform distribution from 0 to 1; spline is a function returning the value of the spline.

3. Calculating the linear extinction coefficients $\mu_{tot}(E)$, $\mu_{foto}(E)$, $\mu_C(E)$, $\mu_K(E)$ and $\mu_{par}(E)$:

$$\begin{array}{ll} \mu_{tot}(E) = \text{spline}(\mathbf{a}_{tot}, \mathbf{b}_{tot}, \mathbf{c}_{tot}, \mathbf{d}_{tot}, E) & \mu_{foto}(E) = \text{spline}(\mathbf{a}_{foto}, \mathbf{b}_{foto}, \mathbf{c}_{foto}, \mathbf{d}_{foto}, E) \\ \mu_C(E) = \text{spline}(\mathbf{a}_C, \mathbf{b}_C, \mathbf{c}_C, \mathbf{d}_C, E) & \mu_{par}(E) = \text{spline}(\mathbf{a}_{par}, \mathbf{b}_{par}, \mathbf{c}_{par}, \mathbf{d}_{par}, E) \end{array}. \tag{5}$$

4. Determining the coordinates of the hypothetical quantum exit from the scintillator as intersection point with the surface S the straight line given by the formula:

$$\begin{cases} x_h = x + a l_h \\ y_h = y + b l_h \\ z_h = z + c l_h \end{cases}, \tag{6}$$

where l_h is the distance from the interaction point (entrance) to the point of photon's hypothetical departure from the scintillator.

5. Drawing a photon run on condition of its interaction before the departure point, correction of the weight factor and the interaction counter:

$$l = -\frac{1}{\mu_{tot}(E)} \ln\left(1 - \xi\left(1 - e^{-\mu_{tot}(E) l_h}\right)\right), \quad w = w\left(1 - e^{-\mu_{tot}(E) l_h}\right), \quad k = k + 1. \tag{7}$$

6. Calculating the coordinates of the interaction point:

$$\begin{cases} x = x + a l \\ y = y + b l \\ z = z + c l \end{cases}. \tag{8}$$

7. Drawing the interaction type

$$\begin{cases} \xi \le \dfrac{\mu_{foto}(E)}{\mu_{tot}(E)} \rightarrow photoeffect \\ \dfrac{\mu_{foto}(E)}{\mu_{tot}(E)} < \xi \le \dfrac{\mu_{foto}(E) + \mu_C(E)}{\mu_{tot}(E)} \rightarrow Compton's\ effect \\ \dfrac{\mu_{foto}(E) + \mu_C(E)}{\mu_{tot}(E)} < \xi \le \dfrac{\mu_{foto}(E) + \mu_C(E) + \mu_{par}(E)}{\mu_{tot}(E)} \rightarrow effect\ of\ pair\ production \\ \xi > \dfrac{\mu_{foto}(E) + \mu_C(E) + \mu_{par}(E)}{\mu_{tot}(E)} \rightarrow Rayleigh's\ effect \end{cases}. \tag{9}$$

8. For the case of photo-effect, go to the step 10.

9. For the case of Rayleigh scattering, drawing the secondary photons direction (the angles θ and φ). The energy of the photons are still the same: $E = E$. Go to the step 14.

10. Drawing the secondary electrons direction with the energy E (the angles θ_e and φ_e). Go to the step 12.

11. For the case of Compton scattering, drawing the azimuthal and polar angles θ and φ for the secondary photons.

12. Drawing the direction and the energy of secondary electrons (the angles θ_e and φ_e). Go to the step 13.

13. For the case of pair creation, drawing the azimuthal and polar angles θ and φ for secondary photons with energy E_γ=611 keV. Doubling photons weight factor: w=2w.

14. Drawing the direction and energy of the secondary electrons (the angles θ_e and φ_e).

15. Calculating the unit vector of the secondary electron direction:

$$\begin{cases} a_e = a\cos\theta_e - (b\sin\varphi_e - ac\cos\varphi_e)\sqrt{\dfrac{1-\cos\theta_e^2}{1-c^2}} \\ b_e = b\cos\theta_e + (a\sin\varphi_e + bc\cos\varphi_e)\sqrt{\dfrac{1-\cos\theta_e^2}{1-c^2}} \\ c_e = c\cos\theta_e - (1-c^2)\cos\varphi_e\sqrt{\dfrac{1-\cos\theta_e^2}{1-c^2}} \end{cases} . \quad (10)$$

16. Determining the point coordinates of hypothetical quantum exit from the scintillator as intersection point with the surface S by the straight line given by the formula:

$$\begin{cases} x_e = x + al_e \\ y_e = y + bl_e \\ z_e = z + cl_e \end{cases} , \quad (11)$$

where l_e is the distance from the photon interaction point (secondary electron creation) to the point of hypothetical secondary electron departure from the scintillator.

17. Calculating the energy transferred to scintillator material by the secondary electron using the continuous delay approximation [15]:

$$E_{eab} = \min\{D(E_e)l_e, E_e\}. \quad (12)$$

18. Increasing the local absorbed energy adder by the registered photon energy: $E_{ab} = E_{ab} + E_{eab}$.

19. For the case of photo-effect or $k > k_{max}$, go to the step 22.

20. Calculating the unit vector of new photon direction via the unit vector of the recent direction (the index *old*)

$$\begin{cases} a = a_{old}\cos\theta - (b_{old}\sin\varphi - a_{old}c_{old}\cos\varphi)\sqrt{\dfrac{1-\cos\theta^2}{1-c_{old}^2}} \\ b = b_{old}\cos\theta + (a_{old}\sin\varphi + b_{old}c_{old}\cos\varphi)\sqrt{\dfrac{1-\cos\theta^2}{1-c_{old}^2}} \\ c = c_{old}\cos\theta - (1-c_{old}^2)\cos\varphi\sqrt{\dfrac{1-\cos\theta^2}{1-c_{old}^2}} \end{cases} . \quad (13)$$

21. Go to the step 4.

22. Increasing the absorbed energy and absorbed energy square adders by the registered photons: $\sum E_{ab} = \sum E_{ab} + E_{eab}$ and $\sum E_{ab}^2 = \sum E_{ab}^2 + E_{eab}^2$.

23. Go to the step 1.

IOP Conf. Series: Materials Science and Engineering **189** (2017) 012035 doi:10.1088/1757-899X/189/1/012035

24. Calculating the mean value, the square of the absorbed energy and the accumulation coefficient of fluctuations:

$$\overline{E_{ab}} = \frac{\sum E_{ab}}{N},\ \overline{E_{ab}^2} = \frac{\sum E_{ab}^2}{N},\ \eta = \frac{\sqrt{\overline{E_{ab}^2}}}{\overline{E_{ab}}}. \quad (14)$$

26. The calculations end.

3. Verification of the Model Results

The results are similar to the data from the work [9-16]. As expected, mean and square values of registered photons absorbed energy are close to zero when scintillation detector thickness or radius approaching zero.

4. Conclusion

The offered method for modeling registration process of high-energy X-ray radiation has high efficiency. It can be used for estimation of X-ray radiation registrar parameters for systems of digital radiography and computer tomography.

5. Acknowledgements

This work was supported by the Tomsk Polytechnic University, project VIU−INK−42/2016.

References

[1] Mathew V 2014 *Quality Magazine* Retrieved from http://www.qualitymag.com/articles/92208-industrial-digital-radiography
[2] Li L *et al* 2016 *Applied Radiation and Isotopes* **114** 188–95 DOI: 10.1016/j.apradiso.2016.05.018
[3] Hamm R W *and* Hamm M E (Eds) 2012 *Industrial accelerators and their applications* (Singapore: World Scientific)
[4] Ueaska M *et al* 2014 *J. of Physics B: Atomic, Molecular and Optical Physics* **47** 234008 DOI: 10.1088/0953-4075/47/23/234008
[5] Li Y *et al* 2003 *IEEE Nuclear Science Symposium Conference Record* **2** 8066819 DOI: 10.1109/NSSMIC.2003.1351949
[6] Chakhlov S V et al 2016 *J. of Physics: Conf. Ser.* **671(1)** 012024 DOI: 10.1088/1742-6596/671/1/012024
[7] Wells K *and* Bradley D A 2012 *Applied Radiation and Isotopes* **70** 1729–1746 DOI: 10.1016/j.apradiso.2012.01.011
[8] Kasianov V A *et al* 2006 *Sensors & Systems* **3** 10−3
[9] Zav'yalkin F M *and* Osipov S P 1985 *At Energ.* **59** 281–3 DOI: 10.1007/BF01123317
[10] Zav'yalkin F M *and* Osipov S P 1986 *At Energ.* **60** 186−8 DOI: 10.1007/BF01371189
[11] Broder D L *et al* (Eds) 1973 *Handbook on radiation protection for engineers* **1** (Moscow: Atomizdat) (In Russian)
[12] Ali E S M *and* Rogers D W O 2011 *Physics in medicine and biology* **57** 31−50 DOI: 10.1088/0031-9155/57/1/31
[13] Schiff L I 1951 *Physical review* **83** 252–3 DOI: 10.1103/PhysRev.83.252
[14] *Gamma data for elements* Retrieved from http://www.ippe.ru/podr/abbn/libr/gdfe.php
[15] *National Nuclear Data Center* Retrieved from http://www.nndc.bnl.gov
[16] Mustafina R M *et al* 2016 *J. of Physics: Conf. Ser.* **671** DOI: 10.1088/1742-6596/671/1/012046